Human-Environment Relations

Emily Brady • Pauline Phemister
Editors

Human-Environment Relations

Transformative Values in Theory and Practice

Springer

Editors
Dr. Emily Brady
Institute of Geography
School of Geosciences
University of Edinburgh
Drummond Street
Edinburgh EH8 9XP
UK
Emily.Brady@ed.ac.uk

Dr. Pauline Phemister
School of Philosophy, Psychology
and Language Sciences
University of Edinburgh
Dugald Stewart Building
3 Charles Street
Edinburgh EH8 9AD
UK
p.phemister@ed.ac.uk

ISBN 978-94-007-2824-0 e-ISBN 978-94-007-2825-7
DOI 10.1007/978-94-007-2825-7
Springer Dordrecht Heidelberg London New York

Library of Congress Control Number: 2012931442

Printed on acid-free paper

Springer is part of Springer Science+Business Media (www.springer.com)

Acknowledgements

We are grateful to the British Academy for funding the workshops on 'Embodied Values' from which this volume derives and to the Institute of Advanced Studies in the Humanities at the University of Edinburgh for hosting the workshops. We would also like to thank all those who took part in the workshops, including those whose contributions could not be included here, and especially Tom Bristow for his invaluable assistance throughout the project.

Contents

Introduction

Emily Brady and Pauline Phemister

The essays in this volume are revised discussions of ideas originally presented in a series of interdisciplinary workshops held over 2 years (2007 and 2008). Under the general heading, 'Embodied Values', workshop participants were invited to question the traditional separation of humans from other animate beings and from the environments they inhabit. They were also asked to explore the notion of a 'space of values' and to consider the possibility of transfers of values within this space not only from humans to their environments, but also in the opposite direction from environments to humans. This work led to a questioning of the very boundaries between the non-human and the human, suggesting more a symbiotic relationship in which the identity and values of the one are informed by the identities and values of the others.

The selection of papers included here represents a range of theoretical and practical approaches to the question of the nature of human-environment relations, focusing not on material exchanges, that is, on the ways in which the physical environment affects and is affected by our physical, chemical and technological interactions, but rather on the ways in which human values and values embedded in the non-human interact and are mutually dependent on each other. Focusing on the notion of environments as spaces and places of moral, aesthetic and spiritual significance, the authors critically examine how external environments transform human values and, conversely, identify concrete ways in which values located in self and society are transferred to environment. Lying at the centre of these essays is the concept of a 'value-space'. Conceiving the value-space as a space in which values move back and forth between environment and humans has the potential, we believe, to lead to a fundamental revision of our conception of the boundaries between the self and its environment. Working with the notion of a more situated self, the essays seek to uncover a certain fluidity in the hitherto set boundaries between the self and nature and a degree of agency in the material world. It is through these ways of rethinking human-environment relations that we see this value-space as having transformative significance.

This value-space is explored through a dialogue of multidisciplinary perspectives from philosophy, literature, theology, human geography and ecological art. These approaches share the common goal of critically advancing knowledge of how human/environment dualism can be both crossed and overcome. Acknowledging discussions that claim that both human and natural values are in some senses culturally constructed, we also recognise that values are exchanged between environment and human in cultivated landscapes, as well as in artistic interventions in the land and in built environments. How do aesthetic and other values emerge within a 'value-space' and become part of the self through embodied practices such as gardening, art-making and fishing? How are landscapes and marine environments imbued with new meaning through human interactions with nature? How is the self shaped by concrete engagements with the rest of nature, including non-human animals? What are the conditions that permit our experiencing external spiritual, aesthetic and moral values of the environment as human values? These are some of the questions raised by the notion of a 'value-space', the conception of which may allow us to formulate new methods to overcome the various dualisms implicit within these questions.

The book comprises two parts: the first addresses theoretical approaches to transformative values, while the second explores applications of theory to practice. In the first part, Alan Holland, Pauline Phemister, Simon James and Patrick Curry consider the metaphysical foundations and philosophical implications of transformative value-space. Rather than being marked by separation and dichotomy, this value-space is characterised by relationality, fluidity, and embodiment. Exploring this space encourages a re-visioning of the role of human and non-human valuers in inhabited places and highlights the need for corresponding modifications to the language of human-nature engagement. Meanwhile, in the latter papers of this first part, Alison Stone, Tom Bristow and David Cooper consider transfers of value within space from the perspectives of aesthetics and cultural and literary theory.

In mainstream environmental philosophy, a preferred way of construing the external and natural world is as a collection of intrinsically valuable items, both human and non-human. On this view, environmental decision-making should be concerned with the maintaining and fostering of intrinsic value. Holland's essay provides an ideal opening to this volume by questioning this dominant discourse. After rehearsing a number of objections to the standard view, he offers an alternative construal of the value-space: one that centres on the concept of 'meaningful relations'. As well as giving an account of the concept and offering a sketch of the value system that forms its backdrop, he suggests that the concept of 'meaningful relations' is a unifying concept that characterises evolutionary and ecological relations as well as cultural ones. On this alternative view, environmental decision-making should be concerned with the continuation of meaning rather than the preservation of intrinsic value.

In her contribution, Pauline Phemister sketches a metaphysical theory of a 'space of value' composed of 'places of value'. She draws on a panpsychist and broadly Leibnizian ontology of living beings as body-soul unities, and outlines a relational theory of teleological or psychical (as opposed to physical) space, according to which both value-space as a whole and individual places of value arise from the perceptual and appetitive relations among the souls and soul-like entelechies of living beings. Co-extensive with the physical space that results from relations among bodies whose interactions accord with the laws of efficient causation, teleological space arises from relations among souls whose operations are subject to the laws of final causation, and is described qualitatively in terms of creatures' pleasure and pain, wellbeing and happiness. Just as physical bodies resist each other within physical space, so too creatures' desires and goals are often opposed. Successful conflict resolution is most likely when rational beings act in accord with the ideal of justice as universal love and wise benevolence, understood (*pace* Leibniz) as encompassing all living things.

Management of conflicting interests is also at the forefront of Simon James' contribution. It is often thought that to conserve nature is simply to prevent certain parts of it from being adversely affected in some way – from being hunted to extinction, say, or paved over and turned into parking lots. However, James – in a move that resonates with Holland's contribution – argues that nature conservation can also involve efforts to safeguard the various meanings nature has for human beings. In making his case, he draws on Martin Heidegger's suggestion that, in the midst of an increasingly technological world, one would do well to cultivate a 'releasement towards things'. James argues that to 'release' oneself towards nature in this way is, amongst other things, to remain open to – or to conserve – the meanings that it has for us. To illustrate what this might involve, he draws on examples of nature writing, from writers such as Richard Mabey and Mark Cocker, and then asks whether nature's meanings ought to be conserved, considering some reasons for thinking that they should.

Moving from meaningful relations to embodiment, Patrick Curry makes a case for the positive revaluing of the body, the feminine, and the Earth as integrally linked aspects of a healthy *modus vivendi* that have been dangerously damaged by modernity. So, in Chap. 4, Curry presents Merleau-Ponty's embodied phenomenology as a promising way to avoid modernity's positionings of mind/spirit as comprising nothing but consciousness and sentience, and of body as nothing but inert matter. For the feminine dimension of this project, he draws on work by Irigaray and Butler, while the indispensability of animist agency is maintained by reference to Plumwood and Abram. From this point, he argues for the centrality of places (as against space) and thence of the Earth, and for an understanding of the Earth that is not just ecological but fully ecocentric.

The separation of human beings from their natural environments has been 'amplified', historically, through drawing a sharp distinction between the significance that art and nature respectively have for us when we appreciate them. To question the sharpness of this distinction is, therefore, to question one aspect of an alleged separation of humans from their environment, and at the same time to consider one 'interface' of culture and nature as a 'space' into which evaluative concerns penetrate. With these considerations in mind, Alison Stone, Tom Bristow and David Cooper examine ways in which cultural, literary and aesthetic experiences open up, rather than close down, human-nature relations.

Like Curry, Alison Stone also tackles the thorny issue of dualism. In Chap. 5, Stone outlines a dilemma: if we resist human/nature dualism by thinking of human beings as embodied and environmentally situated – as natural – then this seems to imply that, as human products, modern environmentally damaging cultures are after all natural. However, claiming that these cultures are alienated from nature seems to return us to belief in human separateness. Stone explores a solution to this dilemma found in the work of Friedrich Hölderlin. He suggests that nature has divided itself into human subjects and the particular, finite, natural objects of which we are conscious and from which we feel separate. Thus both humanity and natural environments are aspects of nature in an expanded sense, as a self-dividing whole. For Hölderlin, then, human estrangement from nature – and by extension contemporary environmental crisis – is a product of nature itself. This has the seemingly unhelpful implication that we human beings neither can nor should attempt to prevent this crisis. However, Stone argues that this quietism is less unhelpful than it might seem. It is motivated by anti-anthropocentrism; it anticipates Heidegger's scepticism about seeking a technological fix; and it allows for a novel justification of environmentalist practices as ways of preparing for possible change in nature's way of being.

Moving to the cultural context of humans, animals and literature, Tom Bristow (Chap. 6) considers how Jonathan Safran Foer's exploration of the language of human-nature engagements allows us to view the value space as a site of the exchange of values that are first located in society, then transferred to environment and finally – and literally – fed back to the individual. Over and above a creative representation of how the environment relates to human values and environmental sensibility, Foer's work discloses unexplored areas of our humanity, outlining moral engagements that highlight the need to develop more meaningful relations between the human and the non-human, relations that are born from a combination of cultural and ecological factors. Foer's exploration focuses on the question of what constitutes *the creaturely* and, on Bristow's reading, features familial relations as a literary value space in which the global and the geopolitical connect and colour idiosyncratic, historically contingent and localised inflections of values. Concentrating on Foer's *Eating Animals* (2009), Bristow explains how Foer employs relations between language, animality and family in order to deconstruct species barrier dualisms and to move beyond the language of intrinsic value, arguing that Foer's text, therefore, highlights the capacities of the imagination to dwell on meaningful ideas of nurturing and of sustainable relations.

In Chap. 7, David Cooper critically explores a thought that has contributed to the perception of a sharp divide between human beings and their natural environments: the thought that there is a radical contrast between the aesthetic appreciation and significance of art and nature respectively. While attempts have been made to soften this contrast, some of these – both 'naturalistic' and 'constructivist' – implausibly reduce one type of appreciation to the other. A useful strategy for questioning the contrast is to consider 'hybrid' places, like gardens, which cannot be uniquely allocated to the category of art/artefact or to that of nature. Having rejected the idea that appreciation of gardens consists of two independent kinds of enjoyment – of art and of nature – Cooper argues that these hybrid places matter to us as symbols or epiphanies of a deep co-dependence between human beings' creative practice and their experience of nature. Traditions of creative practice shape experiences of nature, just as these in turn shape practice. Not even notionally is it possible to envisage a way of human life in which there is not this co-dependence. It follows from this that the alleged divide between the human and the natural receives no support from a proper understanding of the intimacy of art and nature.

The papers comprising Part II of this collection explore a variety of ways in which we might concretise meaningful relations and transformative values in the context of diverse practices and environments. In these, Michael Northcott, Isis Brook, Reiko Goto Collins and Tim Collins, Andrea Nightingale, and Nathalie Blanc discuss attempts to embody human-environment relations through aesthetic practices, as well as through human social and political engagements with the non-human.

In an essay which enlightens discussions of climate change through theological ideas, Northcott considers how the senses and spirit might help us rethink energy issues in terms of human-environment relations. On his view, the refusal of liberal capitalist societies effectively to regulate greenhouse gas emissions reflects the underlying Cartesian assumptions that persist in the culture of these societies. Since the 1950s, Merleau-Ponty, Karl Bohm and others have sought to unsettle the Cartesian legacy by recovering a more embodied and sensual account of human cognition and consciousness. Northcott suggests that this philosophical repair finds significant echoes in the narrativist turn in theology, and conceives both trends as recalling approaches to knowing and action that may be found in the theology of Saint Paul, and in traditional Christian ritualisation of relationships between past, present and future action and being. On his view, the enduring and embodied character of such inter-temporal beliefs and rituals represent a sensitivity to inter-generational relations concerning energy use and climate change which is required if modern perceptions about the consequences of energy use for the planetary future are to change.

In the following chapter (Chap. 9), Isis Brook brings themes of embodiment to bear on the question of place and placelessness. Although there are, undoubtedly, compensatory benefits – such as cosmopolitanism and increasing self-reflection – that mitigate the psychological and social problems of living un-embedded lives in placeless environments, still, the question remains: How can we have culturally enriching and worthwhile lives when the material and social fabric of our situation does little to nurture or sustain the kinds of relationships with each other and with nature that would seem to be a prerequisite for a healthy life? In response, Brook turns to the transformative power of active, purposive engagement with the material realm and proffers the suggestion that simply by making things, by actively engaging in things and, particularly, by mending things, we can rediscover the necessary environmental virtues to reintegrate ourselves into the material fabric of the world. Moreover, she argues, activities such as these have the capacity to transform even the most barren of our contemporary ‘thinned out’ or ‘place-less’ spaces, turning them into enriching places.

From making and mending of objects for use in our everyday lives, we turn to the creation of artworks in the public arena. Environmental artists play a key role in breaking down artefact/nature and human/nature dualisms through ecological, restorative and biogenic creative engagement with natural things. In Chap. 10, Goto Collins and Collins track the evolution of a specific ‘ethical, aesthetic impulse’ in environmental art practice, looking for indications of value exchange between the human and non-human. Describing the 'ethical-aesthetic impulse' as the move by artists towards nature as a context and subject for work that addresses ethical ideas and aesthetic opportunities in relationship to a changing environment, they outline a theoretical framework that is defined by environmental aesthetics and informed by ideas about relational and dialogic aesthetics. This framework informs their case studies of three artworks involving trees and forests – created by Alan Sonfist, Joseph Beuys, and Helen and Newton Harrison. Goto Collins and Collins consider how these artists have contributed to the freedom, health and wellbeing of trees through the range of spatial, social and political means available to them. Through these case studies, they also examine the relationship between experience, imagination and memory in terms of its potential as a pathway to empathic experience – an experience, they argue, that constitutes an important step in the evolution of ethical values. Finally, Goto Collins and Collins describe how they themselves have integrated these ideas into practice in their own project, 'Eden3: The Secret Life of Trees'.

Staying within the public realm, in the next chapter, we move from the world of art to the world of work. Recent research into emotional attachments to nature suggests that these feelings are bound up in people's willingness to protect nature and, more particularly, that the experiences people have in nature help to form a sense of identification with their environment. In the often neglected context of marine environments, Andrea Nightingale builds on this body of work by exploring how fishermen's embodied interactions with the spaces and places of fishing are integral to their understandings of 'nature' and 'fishing'. As fishermen move from their boats, to the pier, and to the meeting room, embodied, emotional and importantly, political transformations in what it means to fish occur. Nightingale insists that these transformations are crucial to how fishers understand and seek to protect (or not) their fishing grounds and the political space within which they fish. Understanding these transformations is especially important in the Scottish context, where, in the wake of increased efforts to devolve fisheries management to fishers' associations, conflicts over how people care and seek to use the resource are profound and contentious. Nightingale hopes that exploring the embodied interactions that produce particular kinds of attachments to 'nature' may allow new channels for fostering cooperation to emerge.

In the final chapter of Part I, David Cooper addresses that most intimate of human-nature engagement – that which occurs in the act of gardening. The final chapter of Part II and the concluding paper of the volume returns to the garden, this time within a socio-political context. As we see in Cooper's essay, gardens can be understood as significant places of interdependence between nature and humans: each supports the other. Individuals and communities shape and nurture nature, while working with nature through gardening can contribute to wellbeing. In this last chapter, Nathalie Blanc explores these relationships through case studies of urban community gardens in Russia and the Netherlands. She argues for the integration of both an ethics of care and aesthetic engagement into human-environment relations. Care ensures the perpetuation of an existing environment with attention – often at a local level – paid to its particularity. Meanwhile, aesthetic engagement can motivate grassroots green action in the context of the more everyday settings of the places we live. Gardening in a community setting supports social justice by creating public ecological spaces for reflection, relaxation and engagement with nature within the urban fabric of cities.

Overall, the essays collected here are indicative of the vast range and incredible diversity of symbiotic transformative value relations between non-human and human nature. We hope that the ideas contained herein will serve as a catalyst, not only for further research, but also for transformations of human consciousness and understandings of self in relation to other, encouraging new ways of thinking ecologically in terms of values and meanings, as well as to more harmonious ways of acting and being in the world.

Contributors

Nathalie Blanc, CNRS Professor and researcher, Université Paris, specialises in aesthetics and environmental issues, especially relationships to animals in cities and the perception of nature in cities. In 2007, she published 'Cockroaches, or Worlds as Images' in *Contemporary Aesthetics* and 'Les subjectivités cosmopolitiques et la question esthétique' in *Émergence des cosmopolitiques et mutation de la pensée aménagiste*. She has also co-edited, with Jacques Lolive, a special issue of the journal *Cosmopolitiques*, 'Esthétique et espace public' (2007).

Nathalie Blanc
UMR LADYSS 7533
Université Paris 7-UFR GHSS
Case 7001, 2 Place Jussieu
75005 Paris, France
nathali.blanc@wanadoo.fr

Emily Brady is Reader in Aesthetics in the Institute of Geography at the University of Edinburgh. Her research interests include aesthetics, environmental ethics, and eighteenth-century philosophy. She is the author of *Aesthetics of the Natural Environment* (2003), and co-editor of *Humans in the Land: The Ethics and Aesthetics of the Cultural Landscape* (2008) and *Aesthetic Concepts: Essays After Sibley* (2001). Her current project is a book on the concept of the sublime.

Emily Brady
Institute of Geography
School of Geosciences
University of Edinburgh
Drummond St
Edinburgh EH8 9JX
Scotland, UK
Emily.Brady@ed.ac.uk

Tom Bristow is a Lecturer in English Literature at the University of New England, Armidale, New South Wales, having previously completed a BA (Hons) and MA (Hons) at the University of Leicester and a PhD at the University of Edinburgh. Tom is interested in the ecological humanities and his current research focuses on theoretical approaches to and the creative practices of literary geography and ecopoetics. He has presented his research findings in America, Australia, Belgium, China, England, Scotland, Spain and Wales.

Tom Bristow
School of Arts (E11), LG38
University of New England
Armidale
NSW 2351
Australia
tom.bristow@une.edu.au

Isis Brook is a philosopher based at Writtle College in the UK. Her doctoral work was on Goethean science (a kind of qualitative approach to nature) and Husserlian phenomenology. More recently her work has been on themes in environmental aesthetics and also education. She has written on place and is particularly interested in researching the relationship between plants and people especially where this relationship can have healing or generally enlivening impacts. Isis has been Managing Editor of the international journal *Environmental Values*.

Isis Brook
Writtle College
Writtle
Chelmsford
Essex CM1 3RR
England, UK
isis.brook@writtle.ac.uk

Tim Collins, PhD, is an artist, researcher and author interested in the changing ideas about our environment, nature and society. He has worked with scientists, technologists, planners and attorneys as he conducts theory in practice experiments working within an art tradition. The fundamental question underpinning his work is how art can and does reframe or act as an interface between nature/culture perception and value. Collins works with artist, Reiko Goto. They are currently working on *Eden3: the Secret Life of Trees*. From 2000–2005, Collins and Goto directed *3Rivers - 2nd Nature*. From 1997–2000, Collins and Goto led the work on *Nine Mile Run*.

Tim Collins
12 1/1 Vinicombe Street
Glasgow G12 8BG
Scotland, UK
timothymartincollins@gmail.com

David E. Cooper is Professor of Philosophy (Emeritus) at Durham University. He has been a Visiting Professor at universities in the USA, Canada, Sri Lanka, Germany, Malta and South Africa. The most recent of his many books are *Meaning* (2003), *Buddhism, Virtue and Environment* (with Simon P. James, 2005), and *A Philosophy of Gardens* (2006).

David E. Cooper
The Bererns
8 New Road
Chatton
Northumberland, NE66 5PU
England, UK
D.E.Cooper@dur.ac.uk

Patrick Curry is Honorary Research Lecturer in the Department of Archaeology and Anthropology, University of Wales Trinity Saint David. He has held lectureships at the universities of Bath Spa and Kent and is the author of *Ecological Ethics: An Introduction*, 2nd edition (2011), among other books and papers.

Patrick Curry
15 Hofland Road
London W14 0LN
England, UK
pmcurry@gn.apc.org

Reiko Goto Collins is an artist whose subjects of inquiry are living things and natural environments, pursued within a new-genre public art practice. Reiko lives in Scotland, and is a researcher with 'On the Edge' Research at Gray's School of Art, Robert Gordon University, Aberdeen Scotland. Her research focuses upon empathetic relationships that emerge through some forms of ecological and environmental art practice. Her current collaborative project with Tim Collins is called *Eden3* and deals with plant response to environmental change. Reiko has been recognized for her achievements working in applied research in art and ecology since 1990.

Reiko Goto Collins
12 1/1 Vinicombe Street
Glasgow G12 8BG
Scotland, UK
r.goto-collins@rgu.ac.uk

Alan Holland is Emeritus Professor of Applied Philosophy at Lancaster University. After an earlier focus on main-stream philosophy his interest latterly has turned to more practical or 'applied' issues. He was founding editor of the journal *Environmental Values*, and recently co-authored a book on this topic with John

O'Neill and Andrew Light. Current topics of research include: 'How to talk about the experiences of animals' and 'The concept of the unnatural'.

Alan Holland
Department of Politics, Philosophy and Religious Studies
County College South
Lancaster University
Lancaster LA1 4YL
England, UK
a.holland@lancaster.ac.uk

Simon P. James is a Senior Lecturer in Philosophy at Durham University. He is the author of *Zen Buddhism and Environmental Ethics* (2004) and *The Presence of Nature: A Study in Phenomenology and Environmental Philosophy* (2009). He is also the co-author, with David E. Cooper, of *Buddhism, Virtue and Environment* (2005).

Simon P. James
Department of Philosophy
Durham University
50 Old Elvet
Durham DH1 3HN
s.p.james@durham.ac.uk

Andrea Nightingale is a Lecturer in Environmental Geography at the University of Edinburgh. She has worked in Scotland (fisheries) and in Nepal (forestry) with a common theoretical focus on subjectivity, common property and environment. Her work explores questions that arise at the social-natural nexus and tries to theorise across that divide.

Andrea Nightingale
Institute of Geography
School of Geosciences
University of Edinburgh
Drummond St
Edinburgh EH8 9JX
Scotland, UK
Andrea.Nightingale@ed.ac.uk

Michael S. Northcott is Professor of Ethics in the School of Divinity in the University of Edinburgh where he has taught since 1989. He was formerly Associate Professor in the Southeast Graduate School of Theology and lecturer at the Seminari Theologi Malaysia, Kuala Lumpur. He is the author of *The Environment and Christian Ethics,* Cambridge University Press, 1996, *An Angel Directs the Storm: Apocalyptic Religion and American Empire* (2004), *A Moral Climate: The Ethics of Global Warming* (2007), and edited with R. J. Berry *Theology After Darwin* (2009) and with Kyle Vanhoutan *Diversity and Dominion: Dialogues in Ecology, Ethics and Theology* (2010).

Michael S. Northcott
School of Divinity
New College
Mound Place
Edinburgh EH1 2LX
Scotland, UK
M.Northcott@ed.ac.uk

Pauline Phemister is Reader in Philosophy at the University of Edinburgh. She is the author of *Leibniz and the Natural World: Activity, Passivity and Corporeal Substances in Leibniz's Philosophy* (2005) and *The Rationalists: Descartes, Spinoza and Leibniz* (2006). Currently, she is working on a book-length project: *Philosophies of Nature: Environmental Philosophy After Leibniz.*

Pauline Phemister
Philosophy, PPLS
Dugald Stewart Building
University of Edinburgh
3, Charles Street
Edinburgh, EH8 9AD
Scotland, UK
p.phemister@ed.ac.uk

Alison Stone is Reader in Philosophy at Lancaster University, UK. She is the author of *Petrified Intelligence: Nature in Hegel's Philosophy* (2004), *Luce Irigaray and the Philosophy of Sexual Difference* (2006), and *An Introduction to Feminist Philosophy* (2007). She has also published on Judith Butler, Early German Romanticism, the Frankfurt School, and the philosophy of nature.

Alison Stone
County South
Lancaster University
Lancaster, LA1 4YL
England, UK
a.stone@lancaster.ac.uk

Part I
Transformative Values in Theory

Chapter 1
The Value Space of Meaningful Relations

Alan Holland

Abstract We can construe the 'value-space' constituted by humans and their environments in a variety of ways. In mainstream environmental philosophy a preferred way of construing this space is as a collection of intrinsically valuable items, both human and non-human. On this view, environmental decision-making should be concerned with the maintaining and fostering of intrinsic value. After rehearsing a number of objections to this view, an alternative construal of the value-space is offered, which centres on the concept of 'meaningful relations'. An account of this concept is given, along with a sketch of the value system that forms its backdrop. 'Meaningful relations', it is argued, is a unifying concept that characterises evolutionary and ecological relations as well as cultural ones. On this alternative view, environmental decision-making should be concerned with the continuation of meaning rather than the preservation of value.

Keywords Value-space • Value • Meaningful relations • Worthwhile life

The aim of this paper is to argue that the value-space constituted by humans and their environments should be structured around the concept of meaningful relations rather than around the concept of intrinsic value – as tends to be the case in mainstream environmental ethics. The contours of the argument, however, are as much metaphysical as ethical. This is because key concepts, such as 'value', or 'meaning', have what we might think of as 'conceptual reverberations': they structure their conceptual environments both semantically and syntactically. If, for example, we

A. Holland (✉)
Department of Politics, Philosophy and Religious Studies, County College South,
Lancaster University, Lancaster LA1 4YL, England, UK
e-mail: a.holland@lancaster.ac.uk

E. Brady and P. Phemister (eds.), *Human-Environment Relations: Transformative Values in Theory and Practice*, DOI 10.1007/978-94-007-2825-7_1,

make the concept of value central, then we make the relation of valuing central to our human dealings with the natural world and in consequence will tend to regard the maintaining of 'intrinsic value' as what environmental decision-making should be all about. By contrast, on one construal of the term, X and Y might be said to stand in a meaningful relation iff. X has meaning for Y, and vice-versa. This (a) brings mutual two-way processes into the picture; (b) casts humans as interactive agents rather than observers; and (c) casts humans as creative in their dealings with the natural world rather than merely responsive.[1]

1.1 Critique of the Value Option

Two questions arise over the practice of making the concept of value central to our thinking about the environment. The first is how far this concept adequately reflects the nature of our relationship with the natural world. The second is how far this concept adequately reflects the nature of the natural world itself.[2] It will be argued that it does neither. But as a preliminary, we need to take on board Robin Attfield's helpful suggestion about how we should construe ascriptions of value to the natural world. For as he rightly points out, something can be valuable without being valued at all. Let us, then, adopt his suggestion that intrinsic value be understood in terms of 'what there is reason to desire, cherish or foster in virtue of the nature of the state or object concerned' (2001: 52; see also 2006: 54).

So far as our relationship with the natural world is concerned, a first objection is that we have as much reason to fear and detest the natural world as to love, cherish or foster it. Or so it might be argued. True, we can find much in nature to intrigue, transport, fascinate and inspire us. But against this we must set J. S. Mill's damning observation that 'In sober truth, nearly all the things which men are hanged or imprisoned for doing to one another are nature's everyday performances' (1874: 28). At the very least we are owed some account of what there is to cherish about the 'protracted tortures' and 'hideous deaths' that are included among these 'performances' (ibid.: 29). Even Bernard Williams's more modulated notion of 'Promethean fear', explained as 'a fear of taking too lightly or inconsiderately our relations to nature', strikes a very different stance from that of fostering or cherishing, for it entails a 'healthy respect' of the kind that we might have for 'treacherous seas' (1992: 67). And whatever we make of Mill's more colourful observations, it is hard to resist the thought that the natural world can be as much a source of sorrow as of joy. The objection, then, is twofold. First, that the concept of value does little to register the great range of relationships that humans enjoy – and as often endure – with the natural world. Second, that the cheerful ascription of value does little to

[1]It is possible to imagine someone mounting the counter-argument that 'meaning talk' can in principle be reduced to, or translated into 'value talk'. But aside from noting the point, no attempt will be made to do justice to such a rejoinder within the scope of the present paper.

[2]I assume throughout a 'secular' worldview, along lines indicated in Holland (2009).

register the part played by the natural world in some of the grief, disappointment, sorrow and failure that are among the basic ingredients of a fully engaged, fully committed and fully vulnerable human life.

A second objection arises if we understand 'intrinsic value' to mean 'non-instrumental value', and if we take this, in turn, to mean the value that nature has independently of human purposes or goals.[3] The problem is especially acute if nature's value is understood as *exclusively* intrinsic or if, conversely and aside from human beings, intrinsic value is supposed to belong exclusively to nature and natural systems. This is because the doctrine casts humans in the role of detached valuer, rather than engaged participant and is blind, therefore, to the values that might be generated as a result of our working with nature, instead of just observing it. In consequence vast domains of value are simply wiped out – all those values associated for example with the work, craft, nurture and care deployed in agriculture, gardening and other activities, and which are exercised in abundance in our creative and interactive dealings with the natural world.

Turning to the characterisation of nature itself that is implied by the ascription of intrinsic value, an initial problem is raised if we understand 'intrinsic value' in another of its senses, to mean the value that an item possesses by virtue of its intrinsic properties. Such an ascription flies in the face of the fact that natural systems are constituted by networks of *extrinsic* relations – relations that members of the system bear to one another. What sense can we make of the invitation to consider the intrinsic properties of the humble wood anemone, devoid of its woodland setting? What matters ecologically is not whether a cheetah is fast, but whether she is fast enough to catch her next meal.[4]

However, returning to a basic notion of intrinsic value as 'value in its own right', how much intrinsic value can we genuinely claim to discern in the natural world? Holmes Rolston and Robin Attfield are among those who see that there might be a difficulty here, and adopt the concept of a 'disvalue' in order to deal with it. Rolston takes the view that although inherent features of the natural world such as predation, parasitism, waste, suffering and death can appear as 'disvalues' from a partial perspective, from a systemic perspective we see that these are prerequisites both for the sustaining of life and the achieving of value: we see, in effect 'the transmuting of disvalue into value' (1992: 275). However, these observations do not demonstrate that there is a balance of value overall, nor, as Attfield admits, do they show that there might not have been more value in the absence of these disvalues. Nevertheless, he himself is prepared to contend that 'a serious basis' exists for claiming that 'parasitism introduces more value than would exist in its absence' because 'the value of the life of parasitic creatures themselves could well exceed the cost to the creatures who are their victims' (2006: 130). In light of the unpredictable course of natural selection, however, it remains wholly unclear how such a cosmic calculation could

[3]See O'Neill (1992) for a classic account of the various senses that may attach to the term 'intrinsic value'.

[4]Partisans of intrinsic value are not insensitive to this point: 'value seeps out into the system', as Rolston (2001: 145) has it.

even be undertaken, let alone how its results could be confirmed. Rolston remarks that 'Death is not the last word; death is transvalued into life' (1992: 275). However, the problem is that the value of the whole life experiment is precisely what is in question.[5] Even supposing that such a calculation was performed, and yielded the result that total disvalue was greater than total value, would this at a stroke destroy our commitment to the natural world? I suggest that it would not – the reason being that this commitment is not based on an appreciation of its value.

Nor is the situation improved if we look to an illustrative basis for the ascription of value – the value that attaches, for example, to a state of 'flourishing'. There are immediately three difficulties with this notion. One is that not all flourishing states are ones that we can or should endorse. O'Neill (1993: 23) gives the example of a flourishing dictatorship. Another and more serious difficulty is that flourishing, along with cognate conditions such as happiness, wellbeing and the like, is a highly competitive good. Hence the protection or promotion of flourishing is a deeply flawed objective. Black-backed gull and puffin are near neighbours on the cliff-top, but to promote the flourishing of the one is immediately to demote the flourishing of the other. A third difficulty is that welfare in the wild is decidedly thin on the ground, and in the natural world at large, flourishing is arguably the exception rather than the rule. Hence the promotion of flourishing rather goes against the grain of nature than accords with it. At the very least, the incidence of flourishing is patchy, and is certainly not guaranteed even if we understand it in the minimal Darwinian sense as involving merely the capacity to survive and reproduce, since the continuity of most species is compatible with extremely low survival rates.

We might hope to evade these difficulties by adopting O'Neill's expanded understanding of the notion of flourishing. For O'Neill 'The question "What class of being can have a good?" is identical with the question "What class of being can be said to flourish"'. And since 'It makes sense to talk of the goods of collective biological entities – colonies, ecosystems and so on', so also does it make sense to talk of the flourishing of such entities (1993: 21–22). The flourishing that we seek to protect and promote, accordingly, could be understood as flourishing on a larger scale – the flourishing of communities, ecosystems and the like. Unfortunately, this move only raises two further problems. One is the difficulty of identifying a flourishing state in a class of entity such as an ecosystem where we take leave of a context that provides us with some notion of a norm. Attfield is correct to insist that 'the notion of flourishing is .. species-dependent' (1987: 41), the reason being that species afford a context within which we can distinguish between more, and less, flourishing states. Death while copulating is the fate that normally awaits the male preying mantis. In most other species, such a fate would be thought inimical to a creature's flourishing. Even supposing that this difficulty were overcome, and that we could be confident in identifying the flourishing states of larger, collective entities, in that event, the former difficulty recurs, of how far we can or should endorse

[5]In what follows, it will be argued that such heroic but un-demonstrable claims about the overall value of the natural world are unnecessary.

the flourishing of such entities. As O'Neill reminds us, the flourishing of a colony of ants 'might .. involve the death of most of its members' (1993: 22). One can only speculate as to what the flourishing of 'Gaia' might involve, but it has to be an open question whether it is necessarily something that we can or should endorse.

1.2 The Importance of Meaning

With one or two notable exceptions, we frequently under-rate the strength of our quest for meaning, both within the span of our own individual lives and in the sequence of events in which our lives are interwoven. In both cases, the more we can uncover a meaningful trajectory of relationships between then and now, and on into the future, the more meaningful our own span becomes. Continuity and contact with the past, and hope for the future, are important both in culture and in nature. This helps to explain, for example, why both nativeness and venerability may be thought of as features warranting protection in the environment. It also helps to explain the significance of biodiversity, understood not as mere species-richness, but in the sense of the promise of speciation. The sense of 'loss' that is in question when we lament the loss of biodiversity is not to be equated with something like the loss of a key; it is a tragic loss, more akin to the loss of a loved one, with whom we have enjoyed the most meaningful of relationships

One of the 'notable exceptions' is the twentieth century existentialist theologian Paul Tillich.[6] Tillich holds that certain sorts of anxiety are constitutive of human existence: the anxiety occasioned by fate and death; the anxiety occasioned by guilt and condemnation; and the anxiety occasioned by emptiness and meaninglessness. He seems even to say that the 'anxiety of meaninglessness' is the greatest of these: 'For the anxiety of meaninglessness undermines what is still unshaken in the anxiety of fate and death and of guilt and condemnation' (1962: 169). Together, these anxieties conspire to create the conditions for despair. He holds, further, that life at its best involves self-affirmation in the face of these anxieties; and this, he says, requires courage (rather than faith).[7]

When he remarks that 'The anxiety of meaninglessness is anxiety about the loss of an ultimate concern' (ibid.: 54), it may seem at first as if he is referring to what is sometimes called 'the meaning of life'. However, closer inspection suggests otherwise: 'Man's being includes his relation to meanings. He is human only by understanding and shaping reality, both his world and himself, according to meanings and values' (ibid.: 57). The examples he provides – 'a belief breaks down .. one is cut off from creative participation .. and so forth (ibid.: 55) – suggest even more strongly

[6]More recently, it is perhaps David Cooper (2003) who has done most to rescue the concept of meaning from the prison of language to which many philosophers would confine it.

[7]Interestingly, he identifies fanaticism – surrender of the self to something trans-individual – as a flawed response to these anxieties.

that he has in mind what we might call 'meanings in life' rather than the meaning of life. This is certainly the notion that we shall be pursuing here.[8]

It is no accident that the work in which Tillich highlights the importance of meaning also reveals his deep admiration for Stoicism. Regarding Stoicism and neo-Stoicism, he writes: 'They are the way in which some of the noblest figures in later antiquity and their followers in modern times have answered the problem of existence and the anxieties of fate and death' (1962: 20). Or again: 'the Stoic has a social and personal courage which is a real alternative to Christian courage' (ibid.: 21). Tillich's high regard for Stoicism is connected with his recognition of Plato's 'thumos' (as the term is used, for example, in *Republic*) as a seminal notion – that aspect of human nature, often translated as 'spirit', that is central to the virtue of courage. That this is no accident stems from the fact that an appreciation of the importance of meaning is certainly assisted by, and might even require, some shift in our governing notions of the good life – away from the prevailing focus on happiness, preference satisfaction and the like, and towards the notion of the worthwhile life. To appreciate the importance of meaning, you need, like Tillich, to see the importance of courage, and the further connection with idealism and aspiration. This connection is forged by the notion of the worthwhile life.

Things that are worthwhile may need to be fought for. This stems from the fact that the concept of what is worthwhile is normative to its core.[9] It concerns what we think is important and how we think things ought to be, and therefore reflects our ideals and our aspirations. All the same, we cannot in all sincerity fight for what we think is important unless we also believe that we stand on firm ground. This requires not knowledge or information but the virtues of wisdom and understanding. With understanding (in one sense the equivalent of 'knowing the meaning') we are led back again to meaning.[10] Just as it requires courage to find meaning, so also does it require meaning to find courage. This, then, is the conceptual terrain in which the concept of meaningful relations achieves its fullest expression.

Something of the same thought is caught by J. Claude Evans when he writes: 'As human beings, we are a part of the natural world in many different ways, ranging from the basic necessity of assimilating energy from the world around us to practices that enable us to relate to the world in ways that feed the soul with the meaning that constitutes a life genuinely worth living' (2005: 193). An earlier anticipation is to be found, perhaps surprisingly, in the work of the early twentieth century

---

[8]By contrast, to discover that there is after all 'a meaning' to life, or that there is such a thing as 'the meaning of life', could be regarded as an extremely depressing prospect indeed.

[9]It should be stressed that normative judgements are here understood to be fully fallible and open to challenge. It is only some subjectivist readings of moral judgement that misguidedly render them immune to challenge – and in the process deprive them of any normative force.

[10]To say all this is to do little more than rehearse arguments to be found in Plato, Aristotle and the Stoics. Plato's stress on 'thumos' has already been noted, while it is Aristotle who draws attention to the key role played by the intellectual virtues of wisdom and understanding. The Stoics, for their part, laid stress on the 'intelligibility' of the cosmos. In order to 'follow nature' – their key maxim – one had to grasp its meaning ('logos').

economist Frank Knight, who astutely observes that 'The authors of great imaginative literature… have never fallen into any such palpable delusion as the belief that men either strive for happiness or expect to be made happy by their striving' – a view that he clarifies in the telling aphorism that 'man .. is an aspiring rather than a desiring being' (1922: 472).

In fact, the connection between meaning and a worthwhile life may be closer still, for it is at least plausible to claim that the living of worthwhile lives requires nothing more, nor less, than the presence of meaningful relationships. One, quite deliberate, implication of this suggestion is to make the ascription of meaning as normative as that of worthwhile life. However, another, equally deliberate but more important implication is to render the requirement for a worthwhile life quite minimal. It is therefore very much at odds with that proposed by Attfield, for whom a worthwhile life is 'one in which states, activities and experiences of intrinsic value offset ones which there is reason to shun' (1987: 66). So, for Attfield, the living of a worthwhile life requires a preponderance of intrinsic value. On the view proposed here, a single meaningful relationship, such as the presence of someone to love, may suffice. So far as the argument of this paper is concerned, this difference makes all the difference. For the consequence is that a worthwhile life may be pursued, and possibly attained, under the most unfavourable of conditions, where the quest for happiness would almost certainly falter, and most especially under conditions in which we cannot be at all certain whether there is a preponderance in the world of value, of flourishing or of wellbeing. In short, it permits the pursuit of a worthwhile life in the world as we have argued it to be.

1.3 Meaningful Connections

At the same time, the notion of meaning connects closely with notions that are hugely important in our dealings with our environments yet are often ignored, such as those of history, place and identity.

In recent work (O'Neill et al. 2008) it has been argued that far too little attention is paid in environmental decision making to considerations involving history, place and identity which, it is claimed, should act to constrain our decisions as to what kind of future is appropriate. The reason is that 'many environmental goods .. are valued as spatio-temporal particulars' (2008: 153). Furthermore, 'particular places matter to both individuals and communities in virtue of embodying their history and cultural identities' (2008: 163). It is reassuring to note that whatever may be the case in environmental decision making, these considerations are far from neglected in certain areas of nature writing, especially in the work of those whom we may refer to as the 'East Anglian School' – for example, Roger Deakin (2007), Richard Mabey (1996, 2000) and Ronald Blythe (1982, 1986), whose spiritual ancestry is to be found in the likes of the Hampshire cleric Gilbert White, and the Northamptonshire poet, John Clare.

Armed with the concept of meaningful relations we can explain why these considerations are important. It is place, i.e. a particular space at a particular time, that provides the spatio-temporal bedrock for any sequence of events, while history is precisely our attempt to make sense of that sequence of events. Among spatio-temporal identities, lives (of individual organisms) are among the most significant sequences that we encounter, though larger identities such as communities are also hugely significant. It is not true, however, that increases in spatial or temporal extent correlate with increased significance. Rather the reverse. Hope, aspiration – and tragedy, require finitude, and a narrative with a beginning and an end. However, finitude is not enough to generate meaningful relations. The earth's moon had a beginning and will have an end, but its relations with earth and other planetary bodies are purely causal. Causal relations underpin meaningful ones but are much broader in scope. Only living items can have meaningful relations in the sense intended, just as only living items can have environments.

The link with environment is direct. Our environments are, precisely, those elements of our surroundings with which we stand in meaningful relations – what David Cooper terms 'fields of significance': 'In calling an environment a field of significance I mean .. that the items within it signify or point to one another, thereby forming a network of meanings' (1992:170). The Pole Star, for example, is no part of the environment of a rabbit, unlike the weasel from which he must escape and the dry stone wall that affords his means of escape. Thus understood, the prime objective of environmental policy making could be re-described as 'the maintaining of environments'. One advantage of this understanding, among many, as Cooper points out, is that it brings animal issues and environmental issues into a common focus. In both areas, bad practice is characterised by 'destruction of the environment' – or in the terms being proposed here, the destruction or impairment of meaningful relations. This characterisation is as true of intensive animal husbandry as it is of intensive logging.

1.4 The Scope of Meaningful Relations

Thus far, no attempt has been made to define the concept of meaningful relations. We have relied, rather, on our own and the reader's intuitions. There is no doubt that our intuitive paradigms for the notion of a meaningful relation are likely to be cultural – involving friendships, community and so forth – and no attempt has been made to deflect attention from such paradigms, as being the readiest and easiest to grasp. We know only too well how marriages can falter when relationships that once were meaningful become merely habitual and routine, and how the life of a community is threatened when a supermarket moves in and begins to drain away the small daily interchanges that are the stuff of people's lives. So too, those facing retirement are likely to acknowledge how their ability to continue to lead worthwhile lives will depend on their ability to sustain and forge meaningful relationships.

However, if it is to perform the task envisaged in this paper, we need to enlarge our understanding of the scope of the concept of 'meaningful relation' very considerably. We need to extend it to include both ecological and evolutionary relationships – in a word, we need to understand it as a truly ‘biocentric’ concept. This move has already been foreshadowed above, where we connected meaningful relations with environment, and both concepts with life.

In defence of the move, we can first register the fact that ‘nature’ itself is a deeply historical concept and is on this account charged with meaning. The point might be granted even if we are considering the universe as a whole, from the time of the ‘big bang’ event, but it might be more readily granted if we narrow our attention to the biosphere, understood as a naturally evolving historical phenomenon. Natural, in the sense of biospherical, relationships, it can be argued, are a paradigm of meaningful relationships both on account of the (past) history invested in them and on account of the (future) history that they foreshadow. They encompass, for example, all those biotic relations that make evolution, speciation and biodiversity possible – the predator–prey relation, parasitism, symbiosis, mutualism, mimicry and a host more. It is for this reason that destruction of the natural world carries with it so much more than the destruction of our means of subsistence. Among the collateral damage is the destruction of meaning. Hence Oliver Rackham (1986) likens the destruction of the natural world to the tearing of leaves from a book. In effect biotic relations are what make life, and any worthwhile life, possible. Just as a happy life is compromised by misery, so is any worthwhile life compromised, including human life, if biotic relations are impoverished. Meaningful relationships therefore can be evolutionary and ecological, as well as cultural. In this extended sense we might say, as a rough approximation, that life cannot be sustained in isolation (from other life), and hence, as an equally rough approximation, that meaningful relations are those than enable life-forms to cluster in a way that is productive of further life.

Nature invests even the briefest of lives with meaning. One can see that this is so by reflecting that if such a life from long ago is found today in fossilised form, it can be a source of endless fascination for the palaeo-biologist. The argument, be it noted, is that the fascination is evidence of the meaning, not its source. From this consideration too, it can be argued that even the briefest of lives is a life worth living. To argue otherwise would seem presumptuous. Paradoxically, the only cases where we should probably yield to the presumption will involve sentient lives in which the extent of suffering might reasonably be judged intolerable. It is only when sentience, and the capacity to suffer, enter the picture that we need to be watchful in our judgements as to what lives are meaningful, and therefore worthwhile.

For certain purposes the point might be better expressed by saying that meaningful relations have evolutionary, ecological and cultural dimensions – and sometimes all three, as is the case with (human) parent/offspring relations. This yields a further reason for extending the concepts of meaningful relations and worthwhile lives beyond the cultural domain. For lives that are worthwhile in the cultural sense also have evolutionary and ecological dimensions that are expressed in biological and physiological characteristics. Consider, for example, the worthwhile life of a pianist. It would surely be impossible to separate out the cultural components of the musical talents that make her life worthwhile from the biological and physiological components that give her the fluency and subtlety that enable her to express her musical talent.

The chief difference between natural and cultural 'meanings' lies in the fact that cultural relationships will tend to be *meaningful to* the parties concerned, whereas natural relationships will tend not to be, or at any rate, not in the same way. David Cooper affords us a useful illustration when he writes: 'Animals, too, dwell in fields of significance; the droppings at the entrance to the tunnel indicate a fox, which signifies a threat to the badger's young, whose squealing expresses hunger, which refers the badger to the berries behind that tree, the scent on which means the recent presence of a fox' (1992:170). Both badger and fox read the signs but cannot, perhaps, be said to judge that they are meaningful. The droppings, the berries and the scent, though parties to the network of meanings, do neither. In the latter case, the ultimate judgement of meaning has to be a human judgement, but this in no way implies that the meaning is a human projection or creation. Some writers are indeed prepared to construe biotic non-humans as themselves 'meaning-makers' – 'entities capable of semiosis and embodied knowledge' (Youatt 2008: 407). The claim is controversial but is for example a possible extrapolation from the tradition of evolutionary epistemology reaching back to Konrad Lorenz (1952). For if all life-forms are problem–solvers then they (literally) 'embody' certain sorts of understanding. Where human interventions serve in effect to create an externally imposed meaning, as in certain forms of transgenic animals and plants, this 'understanding' is displaced and lost. The properties and habits of transgenic creatures – their phenotypical expression – become more predictable (such is the intention, at least), but they thereby become less natural, since the tendency of 'natural born' entities has always been towards the ungovernable and unpredictable; they have the inherent capacity to surprise.

1.5 Policy and Practice

A final point to address is whether the concepts of meaning and worthwhile life are being understood in such an attenuated sense as to render them useless in the context of environmental policy and practice. To this concern, the briefest response must suffice, but note in the first place how the concept of value scarcely distinguishes itself in this field. Despite the fact that one third of the land mass of the planet is given over to agriculture and gardening, which therefore represent what is probably humankind's most extensive and prolonged engagement with the natural world, an ethic dominated by the belief in nature's intrinsic value must be virtually silent on the merits of this engagement. Indeed, if nature's value is supposed to lie in its independence from human aims and goals, then it must follow that what is touched by humans is thereby <u>de</u>valued. Bringing meaning to the fore, on the other hand, we can argue that the quality of these engagements be judged in light of the meaningful relations that they flow from, express and produce. More concretely still, such judgement bears on some of the more heartening developments in recent agricultural practice, such as the organic movement, the return to extensive agriculture, and farmers' markets. The claim is that these developments are good, in part, because they are ways of building and maintaining meaningful relations both with our fellow humans, and with the non-human world.

Among our larger concerns one might cite, for example, the loss of biodiversity and the impacts of genetic modification. Both gain illumination if we see them as reflecting anxieties over the loss of meaning. An appeal to meaningful relations might explain the rather different public receptions afforded information technology and gene technology respectively. On the one hand, the impact of information technologies has been largely to enhance the possibilities for meaningful relationships, particularly in the areas of communication and social activity. However, this could well change, and we are already seeing darker and more destructive uses of IT at work – for example in the service of theft and pornography. On the other hand, for many, GM technology practises forms of displacement and dislocation that disrupt and diminish meaning, especially the meaning invested in 'natural' relationships. This it does, primarily, by overriding meaningful evolutionary and ecological relationships, and substituting molecular ones. GM animals reared in laboratories, in particular, typically suffer this 'loss of environment', much as do conventional domesticated animals reared under intensive systems. Similar points can be made about concerns over the loss of biodiversity, which are regularly misread as concerns about loss of variety and richness – such as genetic engineers might even offer to make good. On the view being proposed here, such concerns are likely to have much more to do with a loss of familiar and meaningful relationships and the dishonouring of a historical legacy – whether it be cultural, ecological or evolutionary. More controversially perhaps is the reflection that we might appeal to meaning as a reason for discontinuing some of the current efforts being made to save endangered species. At least this might be argued in those cases where the prospects for future meaningful relations are poor.

At a smaller scale, and in lighter vein, we can find reason to endorse Roger Deakin's outburst against the practice of flailing hedges: 'I know of nothing uglier or more saddening than a machine-flailed hedge. It speaks of the disdain of nature and craft that still dominates our agriculture' (Deakin 2007). Bringing the concept of meaningful relations to bear, we can reflect on the likely thoughtlessness of the mechanised assault that requires little or no knowledge of the structure and components of the hedge, nor sensitivity to the surroundings. By contrast the hedge-layer, using traditional tools, requires skill and understanding of the subject matter that constitute what Deakin rightly calls a craft. Throughout the process, judgement is called for – what to leave standing, which branches to bend to the floor, where and with what force to apply the billhook. This is interaction of the most rewarding – and meaningful – kind. More contentiously, we can find reason to challenge the increasing resort to nest boxes in bird reserves, a practice which effectively irons out the meaningful distinctions between the nesting habits of different species, so that owls, tits, nuthatches and willow-warblers now misleadingly appear to favour a common environment for the rearing of their young. Or, in more serious vein, we might instance the 2001 outbreak of foot and mouth in the UK. On reading the first-hand accounts of those affected, one is struck by the fact that what is often uppermost is not the material loss – terrible though this was – but the destruction of past histories and the dashing of future hopes. Evidently, historical continuity, together with future prospects that afford room for aspiration, form a major part of what gives meaning to our lives. The concepts of meaning and worthwhile life, therefore, are not only relevant to environmental policy and practice. Their effective deployment can also make a difference.

1.6 Conclusion

The unfinished thought that lies behind the foregoing remarks is simply this: if we are to construct an ethic that is appropriate for the world in which we find ourselves, then we must come to terms with the following two possibilities: (i) that the natural world as a whole may well not exhibit a preponderance of value over disvalue; (ii) that in the natural world at large the flourishing life may well be the exception rather than the rule, even if the concept of flourishing is understood in the most minimal Darwinian sense as indicating the capacity to survive long enough to reproduce. What has been offered is no more than a sketch of what an ethic might look like that eschews the conventional focus on concepts such as value and flourishing. It is based around the concept of meaningful relations and is, arguably, better fitted for the world in which we find ourselves. It recommends that environmental decision-making should concern itself with the continuation of meaning rather than the preservation of value.

References

Attfield, R. (1987). *A theory of value and obligation*. London: Croom Helm.
Attfield, R. (2001). Postmodernism, value and objectivity. *Environmental Values, 10*, 145–162.
Attfield, R. (2006). *Creation, evolution and meaning*. Aldershot: Ashgate.
Blythe, R. (1982). *From the headlands*. London: Chatto & Windus.
Blythe, R. (1986). *Divine landscapes*. Harmondsworth: Viking.
Cooper, D. E. (1992). The idea of environment. In D. E. Cooper & J. A. Palmer (Eds.), *The environment in question* (pp. 165–180). London: Routledge.
Cooper, D. E. (2003). *Meaning*. Chesham: Acumen.
Deakin, R. (2007). *Wildwood: A journey through trees*. London: Hamish Hamilton.
Evans, J. C. (2005). *With respect for nature*. New York: SUNY Press.
Holland, A. (2009). Darwin and the meaning in life. *Environmental Values, 18*, 503–518.
Knight, F. (1922). Ethics and the economic interpretation. *Quarterly Journal of Economics, 36*, 454–481.
Lorenz, K. (1952). *King Solomon's ring: New light on animal ways*. London: Methuen.
Mabey, R. (1996). *Flora Britannica*. London: Sinclair-Stevenson.
Mabey, R. (2000). *Country matters*. London: Pimlico.
Mill, J. S. (1874). Nature. In J. S. Mill, *Three essays on religion*. London: Longmans.
O'Neill, J. (1992). The varieties of intrinsic value. *The Monist, 75*, 119–137.
O'Neill, J. (1993). *Ecology, policy and politics*. London: Routledge.
O'Neill, J., Holland, A., & Light, A. (2008). *Environmental values*. London: Routledge.

Rackham, O. (1986). *The history of the countryside*. London: Dent.
Rolston, H. (1992). Disvalues in nature. *The Monist, 75*, 250–278.
Rolston, H. (2001). Challenges in environmental ethics. In M. E. Zimmerman et al. (Eds.), *Environmental philosophy: From animal rights to radical ecology* (3rd ed., pp. 126–146). Englewood Cliffs: Prentice Hall.
Tillich, P. (1962). *The courage to be*. London: Fontana.
Williams, B. (1992). Must a concern for the environment be centred on human beings? In C. C. W. Taylor (Ed.), *Ethics and the environment* (pp. 60–68). Oxford: Corpus Christi.
Youatt, R. (2008). Counting species: Biopower and the global biodiversity census. *Environmental Values, 17*, 393–417.

Chapter 2
Relational Space and Places of Value*

Pauline Phemister

Abstract Drawing on a Leibnizian panpsychist ontology of living beings comprising both body and soul, this chapter outlines a theory of space based on the perceptual and appetitive relations among these creatures' souls. Co-extensive with physical space founded on relations among bodies subject to efficient causation, teleological space results from relations among souls subject to final causation, and is described qualitatively in terms of creatures' pleasure and pain, wellbeing and happiness. Particular places within this space include the kingdom of grace, where morally responsible, rational beings act as far as possible in accord with the ideal of justice as universal love and wise benevolence. However, while Leibniz considered love as properly directed only towards rational beings, it is argued here that the truly wise person will direct their love and benevolence towards all living things.

Keywords Leibniz • Space • Place • Relations • Love • Justice

My dog and I take an early morning walk on the beach. The tide is out and we head to the firmer still-wet sand. He runs after the birds as they breakfast on worms. I feel the wind in my hair, the movement of my limbs, hear the lapping of the waves and the cry of the gulls. I look at the sun as it rises, spreading soft pink shades up into the lightening sky. I notice the soft, subtle colourings of the pebbles and shells and find a few crabs in the rockpools. I notice the distinctive smell of seaweed and taste the saltiness in the air as, in a light breeze, it comes off the sea. In the evening, I recall my morning walk. Presumably the tide has come in and is again receding. Are the birds congregating before they sleep? What of the fish in the sea, the worms

---

* A longer version of this paper appeared in Phemister (2011).

P. Phemister (✉)
Department of Philosophy, Dugald Stewart Building, University of Edinburgh,
3, Charles Street, Edinburgh EH8 9AD, Scotland, UK
e-mail: p.phemister@ed.ac.uk

E. Brady and P. Phemister (eds.), *Human-Environment Relations: Transformative Values in Theory and Practice*, DOI 10.1007/978-94-007-2825-7_2,

and crabs in the sand and insects in the air? What would the beach be like were no living being experiencing it? Would the sand still be golden and the air salty? Would the seaweed still have its distinctive smell? Or is the beach in itself no more than a mass of atoms, gases, chemical elements, molecules, sub-atomic particles, quarks, electro-magnetic forces and gravitational fields such as physicists and chemists describe? The contemporary physicalist model developed from the separation of the mental and physical whose orthodoxy was established through the efforts of philosophers and scientists, such as René Descartes, Pierre Gassendi, Christiaan Huygens, Robert Boyle and John Locke during the scientific revolution of the early modern period.

The separation appears most starkly in Cartesian metaphysics. For Descartes, the essence of body is extendedness. Every body is an extended thing, *res extensa*. Bodies are portions of matter extended in length, breadth and depth. They are divisible and measurable, amenable to mathematical description. Operating in accordance with strict rules of collision and deterministic laws of nature, they are not free, nor do they think or perceive. Minds, in contrast, think, perceive, desire and sense, but they are not spatially extended, divisible, measurable or subject to physical laws. Instead, they have free will and self-consciousness, and are subject to moral judgment, reward and punishment. No attribute of body is attributable to mind, nor does any attribute of mind pertain to body. From this, Descartes concludes in his *Sixth Meditation* that each is completely independent of the other; neither requires the existence of the other in order to be (1985–1991: II 54).

In this way, Descartes secured the mind's possible immortality: being independent of body, it can survive the dissolution of its body. Moreover, bodies can survive the annihilation of all minds or souls. Minds are conceived, not as parts of nature, but as above or outside nature. The mechanistic natural world does not depend upon anything mind-like. For Descartes, the beach is simply extended body – physical space, variously modified by its moving parts into the sand, sea, rocks, birds and fish that are themselves composed of physical moving particles, indefinitely divisible into increasingly smaller material parts. These parts have motion, size, shape and other so-called 'primary' qualities, but are essentially mere cogs in a physical machine. Although Descartes acknowledges life and sensation in non-human animals, he explains these in purely mechanical terms. Life, he tells Henry More (5 February 1649), is simply the heat of the heart (1985–1991: III 366). Animals are mere mechanisms, he tells the Marquess of Newcastle (23 November 1646), 'responding' to external stimuli in much the same way as does a sleepwalker (1985–1991: III 303).[1] Meanwhile, for Descartes, the colours, tastes and other 'secondary' qualities perceived by humans exist only in minds. Or, if they do exist in bodies, they are there only, as Descartes' near contemporary John Locke proposed, as 'powers' in bodies to produce the ideas and sensations of them in our minds, powers

[1]Tensions and ambiguities in Descartes' account of animal sensation are outlined in Cottingham (1998). See also Clarke (2003: 71–77).

that are reducible to the primary qualities of bodies' constituent parts. Ultimately both Locke and Descartes consider the physical world as describable entirely in material terms, without reference to anything mental or spiritual.

The rise of this mechanical model did not proceed unchallenged even in its infancy. Others, appealing to earlier traditions, resisted the divorce of the material or natural from the spiritual or supra-natural. For instance, neo-platonists Henry More and Ralph Cudworth postulated causally efficacious 'vital principles' (More) and 'plastic natures' (Cudworth) throughout the world of matter. Today, a growing number of panpsychists favour a new understanding of the 'physical' that incorporates the mental in a non-reductive manner, conceiving bodies as capable of experiences or feeling in all their parts.[2]

Adopting a panpsychist metaphysics of nature does not lead inexorably to the holding of benevolent attitudes towards non-human life. Nonetheless, viewing the physical universe as a world full of experiencing, feeling beings certainly encourages empathic affinities with the non-human and makes it harder for us to treat the world with impunity, to regard it solely as a resource for our own gratification or to lack remorse when we do so. Care, we insist, should be taken to minimise suffering wherever it is found. However, this is easy to say and less easy to do. Not knowing from the inside, so to speak, how other beings experience the world, we focus upon their physical features and resort to purely scientific management of them and of the places they inhabit. In what follows, I explore the possibility of conceiving these places, not just as physical places filled by material objects, but also as spiritual places replete with experiencing minds and mind-like things. Conceiving mental dimensions of space may in turn clear a way for a new spatial paradigm to emerge, one that will facilitate co-operation between the natural sciences and the humanities towards a truly ecological management of the environment. To begin, I propose we revisit the scene of seventeenth century dissent to listen anew to one of the voices raised then in opposition to the Cartesian separation of mind and body. That voice belongs to Gottfried Wilhelm Leibniz.

2.1 Leibnizian Panpsychism

Leibniz's anti-Cartesianism touches all aspects of Descartes' thought, but is particularly critical of Descartes' stark separation of mind and body. On Leibniz's view, every soul is joined to an organic body with which it forms an indivisible, unified living creature or 'corporeal substance'.[3] There are 'no *separated souls*, nor

[2]E.g. Strawson (2006).

[3]For 'corporeal substance' interpretations of Leibniz, see Garber (2009), esp. chapter 2; Phemister (2005); Smith and Nachtomy, eds. (2011). The classic statement of the position occurs in Leibniz's letter to De Volder of 20 June 1703 (1989: 177).

spirits [*Génies*] without bodies', he states clearly in the *Monadology* (1989: 222). Moreover, each soul's body is itself a composite of other living things:

> each living body has a dominant entelechy[4], which in the animal is the soul; but the limbs of this living body are full of other living beings, plants, animals, each of which also has its entelechy, or its dominant soul. (Ibid.)

Deprived of dominant entelechies, objects such as rocks, mountains and tables are inanimate. Nevertheless, even they are compounds of living creatures. Within any piece of matter are living beings whose organic bodies are further divided into other living beings that feel and sense the world around them. Thus the beach, on Leibniz's view, is no mere inert mass of lifeless particles. Instead, in all its parts, 'there is a world of creatures, of living beings, of animals, of entelechies, of souls' (1989: 222).

Each mind, soul or entelechy has desires and appetites, and acts always in pursuit of what, often sub-consciously, it perceives as conducive to the preservation of its being. The parasite searches for a hospitable host; the dog seeks a warm, dry comfortable place to sleep. Meanwhile, despite there being no actual interaction between souls and bodies, their bodies move in ways suited to these desires, bringing into being a harmony pre-established by God. Consequently, every event is explicable by both final and efficient causation, giving rise to what Leibniz calls two natural kingdoms, one of efficient causes and the other of final causes. In the former, bodies act and are acted upon in accordance with the laws of motion and resistance; the latter is populated by perceiving, appetitive minds, souls and entelechies, governed by laws of final causes in their pursuit of that which they perceive as good or beneficial and their avoidance of what appears harmful (1989: 223). By way of illustration, Leibniz instances his writing of the *Monadology* (1989: 217). The act of writing involves the movements of his body, as well as an infinite number of other efficient causes (such as those that went into the production of the paper, his writing implements, and so forth). In parallel, however, it also involves the 'infinity of small inclinations and dispositions' of his soul that led to his desire to express his thoughts in print (the final cause).

In deciding to commit his thoughts to paper, Leibniz acted freely. Free agency permits a further distinction, whereby the 'physical kingdom of nature' is contrasted with a 'moral kingdom of grace' or 'divine city of minds' ruled by God as monarch (1989: 224), where self-conscious minds pursue ends they have freely chosen for themselves and for whose outcomes they are held morally responsible. This moral kingdom, it will be suggested below, resides within the wider natural kingdom of final causes that encompasses *all* beings that act by final causes. Thus, the natural world is *both* an extended physical world of bodies and efficient causes *and* a world of perceiving, desiring souls and final causes, some of which are freely chosen as

---

[4]'Entelechy' is the Aristotelian term Leibniz appropriates to signify the 'form' of the body. The terms 'mind', 'soul' and 'entelechy' signal differences of degree (of activity and distinct perception) rather than of kind.

personal goals by self-conscious minds. With regard to bodies, Leibniz proposed a relational theory of physical space founded upon bodies' relational qualities.[5] It is to this that we now turn.

2.2 Extended Relational Space and Places

Leibniz defended his relational theory of physical space in the course of a brief correspondence with the Newtonian theologian, Samuel Clarke. Actual space is 'nothing else but an order of the existence of things, observed as existing together' (1956: 63).[6] Considered apart from the co-existing things, spatial relations are neither things (subjects) nor qualities (accidents) of things. The relation is not a third thing in addition to the items related. Consequently, there is no body-independent space, no Newtonian absolute and possibly empty space. Space abstracted from bodies is a purely ideal theoretical construction, and is not real (1956: 64, 71).

However, the co-existing things themselves are real and possess relational qualities that ground the ideal spatial relations between them. Thus, a line L has the relational quality of being longer than line M. On the other side, line M has the relational quality of being shorter than L. The ratio between the lines can be stated as 3:1 from the side of L or as 1:3 from the side of M. In both cases, the relation belongs to a subject (1956: 71). When co-existing things stand thus in relation to each other, the space they inhabit is real: space that is 'filled up' is actual space (1956: 63). Real space requires individual existing things whose relational qualities underpin the order of co-existence that constitutes space.

Within this space, we can identify particular places: space, Leibniz tells Clarke, is composed of 'places taken together' (1956: 70). Strictly, however, no part of space remains the same from one moment to the next, for the relations among co-existing things are in constant flux. How then can we speak of places at all and how can we be said to return to the same place? The answer lies in the fact that the changes among co-existing things occur at different rates – some less pronounced or slower than others. Because of this, we are able to pretend that '… among those co-existents, there is a sufficient number of them, which have undergone no change'. These 'fixed existents' serve as anchors for locating places amid the more noticeable changing relations of others. For instance, the relative stability of buildings and natural features of the landscape enables us to identify specific places that other things (whose relations to the fixed existents are more rapid) may enter or leave and to speak as if 'those which have such a relation to those fixed existents, as others had to them before, have now the *same place* which those others had' (1956: 69).

---

[5]Idealist interpretations of Leibniz (e.g. Adams 1994; Rescher 1979) conceive spatial relations as founded upon monads' perceptions.

[6]For detailed analysis of Leibniz's theory, see Vailati (1997: chapter 4).

The space described here is quantifiable or measurable. It is an ordering of co-existing bodies that have size, shape, resistance and motion. It is space filled with material objects whose interactions are explicable mechanistically through laws of efficient causation. However, as we have seen, Leibniz conceives the natural world not just in terms of bodies governed by laws of efficient causation, but also in terms of minds, souls and entelechies governed by final causes. Like bodies, they too co-exist and possess qualities relating them to each other. Is it then possible to describe a parallel *psychical* space as an ordering of co-existing perceivers with their various appetitions, impulsions, desires and volitions? Can the qualitative perceptual and conative qualities of biological entities provide a way of conceiving space, not just as the amalgamation of places filled by physical things, but also as composed of places imbued with souls and their values and goals? The implications of such a conceptual revision are wide-ranging, but my goal here is merely to sketch in broad outline the possible contours of such a space.

2.3 Perceptions and Appetitions

The first step is to examine the psychical qualities from which such a relational space might be constructed. Perceptions (sensations, imaginations, thoughts, feelings of pleasure and pain, emotions and passions) and appetitions (appetites, desires, volitions) are the internal qualities that characterise minds, souls and entelechies. All are inherently relational. Perceptions are always perceptions of something; appetitions are always focused on a desired object or possible state of affairs. Through its perceptions, each living being, Leibniz states in the *Discourse on Metaphysics*, is like a mirror, representing the whole universe (1989: 42). Hidden deep in our perceptual states are representations even of events on Jupiter and Saturn, in the ocean beds and among the minute particles of the earth. Everything is registered at some level in each mental state of every living thing. I perceive the gull flying past and it perceives me. I subconsciously register the tiny organisms within each grain of sand and they in turn insensibly perceive me. As David Abram notes in a discussion of Merleau-Ponty,

> To touch the coarse skin of a tree is thus, at the same time, to experience one's own tactility, to feel oneself touched *by* the tree. And to see the world is also, at the same time, to experience oneself as visible, to feel oneself *seen*. (Abram 1996: 68)

Our perceptual representations of the world differ according to their degrees of distinctness and confusion. Mostly, we perceive things confusedly or even insensibly, failing to distinguish clearly the minute living parts from which all bodies are ultimately composed. When we perceive distinctly, we recognise not only that we are ourselves perceiving, appetitive beings, but also that these features are present in the things perceived. When I perceive the gull distinctly, I know that I perceive it, that it perceives me, and that it has its own desires and needs.

Whether perceptions are distinct or confused depends on the appetitive force responsible for the transition from one perception to the next.[7] The greater the force, the more distinctly the universe will be represented in the ensuing perception, as Leibniz explains in *On the Method of Distinguishing Real from Imaginary Phenomena* (1969: 365). Indirectly, appetition is also responsible for our attitudes towards what we perceive, determining, for instance, whether we regard them with compassion, care and love or with disdain, indifference or hatred. The reason lies in connections between perceptions of perfection, order and beauty and feelings of joy, happiness, love and the desire for justice.

In the distinct perception of nature's diversity, order, interconnectedness and sheer magnificence, Leibniz discerns the source of human joy. In the *New Essays*, he highlights the pleasure we gain from the appreciation of the order and harmony of the world. The joy we experience consists in a sense of perfection. Pain, on the other hand, is a sense of imperfection (1982: 194).[8] We take pleasure from our perception of the beauty and order of the universe and are pained when we perceive its disorder or destruction. Expanding these ideas in *On Wisdom*, Leibniz observes how pleasurable feelings of perfection are invoked through the awareness of perfections such as understanding, courage and beauty, whether in ourselves or in other human beings. Beauty can also be perceived in animals, artworks, handicrafts and so forth. Moreover, 'the image of perfection in others, impressed on us, causes some of this perfection to be implanted and aroused within ourselves', though this is sometimes impaired by our own feelings of jealousy, fear or shame (1969: 425).

Leibniz takes up discussion of the relation of happiness and love in the *New Essays*. Single pleasures, he insists, are merely steps on the way to happiness (1982: 187–9, 194). Happiness is a 'lasting pleasure' (1982: 194),[9] understood not as a permanently pleasurable state, but as a continual progression towards new pleasures (1982: 194).[10] Reason helps us perceive the true good, maintain the generally upward momentum towards new pleasures, and avoid setbacks incurred through infelicitous choices. Hence, 'reason and will … lead us to towards happiness, whereas sensibility and appetite lead us only towards pleasure' (ibid.). Consequently, happiness eludes non-rational beings, who are guided by appetite rather than rational will and incapable of appreciating the rational order and harmony of beautiful things.

'Love', Leibniz defines as a disposition 'to take pleasure in the perfection, wellbeing and happiness of the object of one's love', noting that it 'involves not thinking or asking for any pleasure of one's own except what one can get from the happiness or pleasure of the loved one' (1982: 163). He accepts that his definition entails that feelings of love can be directed only towards rational beings, for only they can experience the happiness and joy that accompany the perception of beauty.

[7] *Monadology* (1989: 215).

[8] See also Leibniz's letter to Arnauld, early November 1671 (1969: 150).

[9] See also 1982: 90 and *On Wisdom* (1969: 425).

[10] Also, *Principles of Nature and Grace* (1989: 213).

> [W]hatever is incapable of pleasure or of happiness is not strictly an object of love; our enjoyment of things of that nature is not love of them, unless by a kind of personifying, as though we fancied that they could themselves enjoy their perfection. (ibid.)

Leibniz believes that people can only truly feel love for non-human animals if they consider animals as rational to some extent.[11] However, this seems unduly restrictive. Whether animals are rational or not, they have appetites, feelings and sensations and are capable of experiencing pleasure and pain. These capacities ground our desire for their wellbeing. We experience disinterested pleasure in seeing them thrive and in seeing their needs met. If this is enough to count as 'love' – and I believe that it is – a revision of Leibniz's definition is warranted, one that does not require animals to be rational in order to be loved. In short, we may drop Leibniz's reference to the happiness of the other, substituting the other's 'specific good',[12] so that love is defined simply as the taking of pleasure in the 'perfection, wellbeing and specific good of the other'.

This revised definition accommodates feelings of love towards animals. Moreover, although few would admit to feelings of love towards minute organisms, it allows for the possibility that unselfish love might be directed towards creatures further down the chain of being – to fish, plants, and indeed to all the living organisms and micro-organisms that Leibniz and other panpsychists conceive in even the smallest parts of matter. These creatures' felt experiences may only echo faintly the pain and pleasure experienced by self-conscious beings, but still one may imagine that some of their feelings tend more towards the pleasant than do others, that they thrive in some environments better than in others and that there is something that counts as their perfection, wellbeing, and specific good. Love towards *all* living creatures can be manifested in the desire to promote their wellbeing, motivated, not by utilitarian concerns, but simply by the pleasure we gain from seeing them thrive.[13]

Love and justice are inseparable: 'Justice', Leibniz remarks in the *Elements of Natural Law*, 'demands that we seek the good of others in itself, and since to seek the good of others in itself is to love them, it follows that love is of the nature of justice' (1969: 137).[14] In his *Diplomatic Codex on the Right of Nations*, he identifies three degrees of justice or 'natural right'. The first two are secular. Commutative justice or 'strict right' stipulates that, as far as possible, no one is to be harmed. Its aim is the preservation of peace, but it does not demand that we positively promote the good of others. Distributive justice or 'equity' obliges us to treat everyone fairly: 'it commands us to *give each one his due*' (1969: 422). The third, and highest, degree of justice – universal justice or 'piety' – operates under God's laws and commands

[11] See *Elements of Natural Law* (1969: 137). In the *New Essays* Leibniz declares animals incapable of giving *reasons* for things, although he allows they can reason inductively (1982: 143). He grants they feel affection and anger (1982: 93, 167), but not misery (1985: 280–281). He was known to endorse vivisection (Smith 2007: 144; also Smith 2011: chapter 1, section 8).

[12] See Aristotle, *Nicomachean Ethics* VI, 7 and X, 5.

[13] This is not to imply that our love for animals is reciprocated. Loving another presupposes the capacity to recognise their perfection, wellbeing and specific good.

[14] See also Riley (1996: esp. chapter 4).

that we live honourably, virtuously or piously, acting always for the good of all. As Leibniz explains in the *New Essays*, it requires that we aim to 'acquire the custom of acting in conformity with reason which makes virtue a pleasure and second nature to us' (1982: 189).

Leibniz often describes universal justice as the 'charity of the wise' or 'universal benevolence'.[15] '[C[haritable people', he explains in the *New Essays*, 'love their neighbours with some measure of tenderness, they are sensitive to the good or harm of others' (1982: 215). In the wise, this love is universal, which for Leibniz means that it is extended to all rational agents. The wise seek universal harmony and the perfection of all individuals and, although no finite being is perfectly wise, the ideal provides a useful guide (Hostler 1975: 53). However, the wise person loves God above all,[16] finding the highest pleasure in the contemplation of the perfect being.[17]

In loving God, the wise seek to bring about God's will, namely universal harmony. From this, Leibniz concludes, in a letter to Arnauld (November 1671), that 'it is the same to love truly or to be wise, and to love God above all things; this is to love all or to be just' (1969: 150). Having revised Leibniz's definition of love to accommodate love towards all living things, we may expect that at the highest level of justice, love is extended beyond humanity to every living thing in creation, in pursuit of true universal harmony. Accordingly, we may also revise Leibniz's remark to Arnauld to read: 'it is the same to love truly or to be wise, and to love God above all things; this is to love all *living things* or to be just'.

2.4 Value Relational Space

Physical space comes into being when bodies stand in relation to each other. So too, however, psychical space – a relational space of value co-extensive with the relational space of bodies – comes into being when minds, souls and entelechies exist in relation to one another. Just as there is no extended empty space before there are physical bodies, so too, in its psychical dimensions, space contains no just, compassionate or loving places until there are actual perceiving beings who act justly and with compassion and love. It contains no places of safety until there are beings relating to each other in ways that are supportive and health-promoting, nor places of terror unless there are belligerent beings intent on each others' destruction. In themselves, love and justice are abstract concepts, ideal and unreal. What are real are the living beings whose relational qualities – their perceptions and appetitions – determine whether the places they inhabit are grounded in relations based in love, kindness and openness or in hatred, envy and secrecy.

[15]E.g. *Diplomatic Codex* (1969: 421) and Leibniz's letter to Arnauld, 23 March 1690 (1969: 360).

[16]*Diplomatic Codex* (1969: 422); *Elements of Natural Law* (1969: 134); *Principles of Nature and Grace* (1989: 212).

[17]*Elements of Natural Law* (1969: 134); *Principles of Nature and Grace* (1989: 212).

Mechanism governs relations among bodies; teleology rules relations among entelechies, souls and minds. Creatures' appetitions impel them towards whatever they perceive (not always correctly) as being in their own best interests. However, just as a body moving into a particular spot must displace any other bodies already occupying that place, so too, the desires and appetites of living things often conflict with one another, making impossible the simultaneous satisfaction of both. The more powerful wins to the detriment of the other. If the eagle is to be fed, the mouse is sacrificed; but the mouse desires to preserve its life and if it succeeds, the eagle goes hungry. Were justice applicable in cases like these, the eagle-mouse relation would conform to that of 'strict justice', whereby as far as possible, no harm is to be done to others. The eagle needs to eat and would not harm the mouse were it not hungry. Unwittingly, the animal kingdom generally operates in accordance with this rule. As Freya Mathews notes, '[a]nimals do not follow the so-called law of 'dog eat dog' (Mathews 1991: 157). Humans, however, often extend their self-interest beyond what is acceptable. We forget to proceed to the next level of justice, justice as equity, which requires us to treat people and other life-forms fairly and to recognise that their right to life may overrule our human desires for material wealth, power, leisure and comfort. The wise apply the highest level of justice, following the principle of universal benevolence, but even they cannot always bring good to all. Nonetheless, as Leibniz informs Arnauld (November 1671), the wise man, 'necessarily strives to please all, even when he cannot do so, much as a stone strives to fall even when it is suspended' (1969: 150).[18]

Competing and non-competing desires and accompanying perceptions are the relational qualities of minds, souls and entelechies from which a space of values comes into being. They determine the nature of this space as individuals react to each other in line with their perceptions of them as beneficial or harmful, as desirable or undesirable, although only rational minds can freely choose to conduct their relations to others in a spirit of universal benevolent love, justice and charity or of hatred, fear and competitive self-interest. Competition frequently occurs within groups of individuals and between nation states and communities, including eco-communities. Modelled upon extended *places* within extended relational space, such groups and communities may be understood as *places* of value within the universal space of values. We consider these in the next section.

2.5 Relational Places of Value

While a corporeal substance's body is located in a particular physical place, its soul or entelechy is situated in a particular place of value, identifiable by the perceptual and appetitive relations in which it stands to others. Consequently, every extended physical place may also be considered as a place of values, describable in terms of the

[18] Leibniz advises that cases of conflict be resolved in favour of the 'better man, that is, the one who loves more generally', for he will use his advantage to help others (1969: 150).

desires, interests, attitudes and needs of the related living beings whose organic bodies in relation comprise that same place conceived as spatially extended. The beach is not just a physical place of related material objects; it is also a place of life – a place where the birds perceive, the constitutive parts of each grain of sand have some bare feeling, and where the flora and fauna perish or thrive. All communities of living individuals constitute places of value, whether this be a rainforest with its incredible diversity of life forms – from termites, insects, birds and primates to all manner of trees and plants – or a human town or city with its different, but equally diverse, conjunctions of species. The particular goals and desires of the individual living things within each community need not always coincide with those of the others – each pursues its own goals – but nonetheless, each community has an identifiable structure and order, the survival of which the majority of its inhabitants desire, even if only unconsciously. The structure and order allows us to speak of such places as eco-communities, that is, as habitats within which individuals cooperate and compete to further their own goals and outside of which these goals could not be realised.[19] Indeed, some members may discover that their goals cannot be realised in a particular environment, perhaps because of pollution, changing environmental factors, or because they have strayed from their natural habitat. Some will migrate; others will adapt to new conditions, rather as a square peg may be hewn in order to fit a round hole; while others will simply not survive. Nevertheless, these places themselves, like their physical counterparts, may be considered relatively stable, despite constant changes in their constituents and their relations, so long as most adhere to a set of common goals. Just as the place in the park is considered to be the same place despite the removal of a bench, so too, the medical practitioners, administrators and patients working together in pursuit of healing permit the construction of the notion of a hospital as a place of value, despite staff and patient turnover.

However, although every physical place is also a place of value, the converse does not hold. Sometimes common goals, shared cultural, ethical and spiritual values establish communities and institutions that are not confined to any particular physical place. For instance, while current staff and students are based on the university campus, alumni continue to belong to the university community even if they live and work far afield. Nor is the teleological place that is the ‘kingdom of grace’ limited to any specific physical place. Those who pursue final ends or goals self-consciously acknowledged as their own, rationally assessing various courses of action before choosing that which they perceive as the best, comprise the ‘moral kingdom of grace’ that Leibniz sometimes, following Augustine, calls the ‘City of God’ – a place of ethical and spiritual values where moral law dictates how each should best behave towards others.[20]

---

[19]Leibniz characterises societies as equal (e.g. among friends) or unequal (e.g. between monarch and subjects) and as limited (constituted for particular purposes) or unlimited (concerned with the ‘common good’). See *On Natural Law* (1969: 429).

[20]Leibniz envisages that minds constitute a society in which God, as ‘lord or monarch’ administers laws of justice that ensure the appropriate reward and punishment of good and evil acts respectively (1989: 67). See also, *Principles of Nature and Grace* (1989: 212); *Monadology* (1989: 224).

Members of God's City are expected to promote the greatest possible happiness of all its citizens, guided by love and the highest form of justice, the charity of the wise. In loving others, we unselfishly desire their happiness, taking for ourselves only the pleasure derived from perceiving the others' perfection. However, in this way others' good is combined with our own, for virtuous action, disinterestedly pursuing the good of all, also constitutes our own perfection.[21] The true good of the wise is inseparable from that of those whom the wise love.[22] Besides, as Leibniz notes in *Reflections on the Common Concept of Justice*, in acting virtuously, the wise attain a serenity of spirit, an internal harmony (1969: 569–570) that mirrors the external harmony of a society in which the good and perfection of all citizens is paramount.

All the same, the City of God forms only part of the natural kingdom of final causes. As physical places comprise extended space, so too places of value comprise psychical space. The kingdom of grace is a subset of the space of values and final causes. Members of the kingdom of grace also belong to the wider space that results from the relational qualities of all living souls and entelechies. Their citizenship of the City of God subjects them to moral scrutiny and appropriate praise or condemnation for acts they self-consciously and freely choose, but it does not follow that such praise or blame should be restricted to acts in relation to other members of the City of God. Given our earlier modifications to Leibniz's definitions of love and justice, the wise are expected to consider the perfection of God's creation as a whole, not just that part that is the kingdom of grace. The wise must take into account both the happiness of human beings and the wellbeing and interests of all other living beings in the space of final causes. In promoting the perfection and goodness of the world in a truly universal sense, the sage will undoubtedly gain immense pleasure from the perception of the unfolding beauty of God's creation.[23] Wherever they may be, whether at the beach, the hospital, the university, the city, the home or the wilderness, the interests of members of the City of God are best served by cultivating attitudes of love and justice, care and consideration towards, and seeking to promote the interests and needs of, all life forms in that place. In the *Elements of Natural Law*, Leibniz observes that 'the science of the just and the useful, that is, of the public good and of their own private good, are mutually tied up in each other, and [that] no one can be truly happy in the midst of miserable people (1969: 132), but this too demands revision so as to read: 'no one can be truly happy in the midst of any suffering whatsoever, irrespective of which life forms are affected'.

[21]*Remarks on the Three Volumes Entitled Characteristics of Men, Manners, Opinions, Times* (1969: 630).

[22]For discussion, see Rateau (2008: 85–91). As Leibniz recognises in *Elements of Natural Law*, wise love often entails some cost, though he still insists that 'even though our own pain intervenes, nothing prevents our taking pleasure in an expectation of the good of others' (1969: 137).

[23]Cp. *On Wisdom* (1969: 428).

2.6 Conclusion

The relational space and places of value described here allow us to conceive the natural world, not just as a world of physical, moving bodies, but also as a world of living perceiving things, whose appetites strive to promote their own wellbeing, and some of whom inhabit the 'smaller' kingdom of grace. The free, self-conscious beings in the latter must assume responsibility not only for those who inhabit the kingdom of grace, but also for others throughout the space of final causes. Space and its component places are formed in part by our perceptions of, and attitudes towards, the living beings from which it results. As humans, we can decide consciously to change our value environments as easily as we change our physical surroundings. The attitudes we adopt towards other living, perceiving, appetitive beings around us alter the tenor, quality and structure of these places and ultimately affect the quality of the value space as a whole.

References

Abram, D. (1996). *The spell of the sensuous: Perception and language in a more-than-human world*. New York: Random House.

Adams, R. M. (1994). *Leibniz: Determinist, theist, idealist*. Oxford: Oxford University Press.

Clarke, D. M. (2003). *Descartes's theory of mind*. Oxford: Clarendon.

Cottingham, J. (1998). Descartes' treatment of animals. In J. Cottingham (Ed.), *Descartes* (pp. 225–233). Oxford: Oxford University Press.

Descartes, R. (1985–1991). *The philosophical writings of Descartes* (3 Vols; J. Cottingham, R. Stoothoff & D. Murdoch, Ed. & Trans.; Vol 3 with A. Kenny). Cambridge: Cambridge University Press.

Garber, D. (2009). *Leibniz: Body, substance, monad*. Oxford: Oxford University Press.

Hostler, J. (1975). *Leibniz's moral philosophy*. New York: Barnes and Noble.

Leibniz, G. W. (1956). *The Leibniz-Clarke correspondence* (H. T. Alexander, Trans.). Manchester: Manchester University Press.

Leibniz, G. W. (1969). *Philosophical papers and letters* (L. E. Loemker, Ed. & Trans.; 2nd ed.). Dordrecht: D. Reidel Publishing Company.

Leibniz, G. W. (1982). *New essays on human understanding*. Cambridge: Cambridge University Press.

Leibniz, G. W. (1985). *Theodicy: Essays on the goodness of God, the freedom of man, and the origin of evil* (E. M. Huggard, Trans.). La Salle: Open Court.

Leibniz, G. W. (1989). *Philosophical essays* (R. Ariew & D. Garber, Ed. & Trans.). Indianapolis and Cambridge: Hackett Publishing Company.

Mathews, F. (1991). *The ecological self*. London: Routledge.

Phemister, P. (2005). *Leibniz and the natural world: Activity, passivity and corporeal substances in Leibniz's philosophy*. Dordrecht: Springer.

Phemister, P. (2011). Relational space and places of value. *Logical Analysis and the History of Philosophy, 14*, 89–106.

Rateau, P. (2008). *La question du mal chez Leibniz: Fondements et élaboration de la Théodicée*. Paris: Honoré Champion.

Rescher, N. (1979). *Leibniz: An introduction to his philosophy*. Oxford: Basil Blackwell.

Riley, P. (1996). *Leibniz' universal jurisprudence: Justice as the charity of the wise*. Harvard: Harvard University Press.

Smith, J. E. H. (2007). The body-machine in Leibniz's early physiological and medical writings: A selection of texts with commentary. *The Leibniz Review, 17*, 141–179.
Smith, J. E. H. (2011). *Divine machines: Leibniz and the sciences of life*. Princeton: Princeton University Press.
Smith, J. E. H., & Nachtomy, O. (Eds.). (2011). *Machines of nature and corporeal substances in Leibniz*. Dordrecht: Springer.
Strawson, G. (2006). Realistic monism: Why physicalism entails panpsychism. In A. Freeman (Ed.), *Consciousness and its place in nature: Does physicalism entail panpsychism* (pp. 3–31). Exeter: Imprint Academic.
Vailati, E. (1997). *Leibniz & Clarke: A study of their correspondence*. Oxford: Oxford University Press.

Chapter 3
Conserving Nature's Meanings

Simon P. James

Abstract It is usually thought that to conserve nature is simply to prevent certain parts of it from being adversely affected in some way – from being hunted to extinction, say, or paved over and turned into parking lots. In the following, however, I argue that nature conservation can also involve efforts to safeguard the various meanings nature has for us. In making my case, I draw on Martin Heidegger's suggestion that, in the midst of an increasingly technological world, one would do well to cultivate a 'releasement towards things'. I argue that to 'release' oneself towards nature in this way is, amongst other things, to remain open to – or to conserve - the various meanings that it has for us, and to illustrate what this involves, I look to some examples of nature writing, from writers such as Richard Mabey and Mark Cocker. In the final section, I ask whether nature's meanings ought to be conserved, and I consider some reasons for thinking that they should.

Keywords Conservation • Meaning • Heidegger • Technology • Nature writing

What does it mean to conserve nature? A distinction is sometimes drawn between conservation (which is thought to involve the judicious or 'wise' use of natural resources) and preservation (which is taken to mean defending nature from human influence). In the following, by contrast, I will suppose that to conserve nature is simply to protect it, both those parts that are thought to be merely useful for human beings and those that are deemed to have non-instrumental value.

It may seem obvious what it means to conserve nature in this sense. It is to ensure that certain vulnerable parts of the world – certain species, for instance, or certain

S.P. James (✉)
Department of Philosophy, Durham University, 50 Old Elvet, Durham DH1 3HN, UK
e-mail: s.p.james@durham.ac.uk

E. Brady and P. Phemister (eds.), *Human-Environment Relations: Transformative Values in Theory and Practice*, DOI 10.1007/978-94-007-2825-7_3,

natural habitats – are not harmed, eradicated or in some other way adversely affected. It is to protect them from being hunted to extinction, or chopped down and burnt, or turned into parking lots. It is to do what organisations like Greenpeace and the World Wide Fund for Nature try to do. My aim in this chapter, however, is to show that nature conservation can take another form; that it can involve efforts to safeguard the various meanings nature has for us. And so I will try to show that, even if our efforts to prevent organisms, habitats, species, etc. from being harmed or eradicated were entirely successful, the work of the nature conservationist would not necessarily be done. Nature's meanings could still require conserving.

To develop my case, I draw upon Martin Heidegger's writings on 'releasement' (*Gelassenheit*) and 'technology' (*Technik*). I do this for two reasons: first, because Heidegger's account of releasement can help one to see what meaning-conservation involves; second, because his account of technology shows why conservation of this sort is needed. In the first section, I explain what Heidegger means by technology, paying special attention to the links he sees between it and what he calls the 'devastation of the earth' (1976: 30). I move on, in Sect. 3.2, to consider his recommendation that, in the midst of our increasingly technological world, we late moderns would do well to cultivate a 'releasement towards things'. In Sect. 3.3, I argue that to 'release' oneself towards natural things is, in effect, to conserve the variety of non-technological meanings nature has for us. To illustrate what this sort of conservation involves, I examine some examples of nature writing from writers such as Richard Mabey and Mark Cocker. My aim in the first three sections is therefore to show that one kind of nature conservation is concerned with safeguarding nature's meanings. In the fourth and final section, I address the difficult question of whether we *ought* to conserve nature's meanings, and I suggest that, with certain qualifications, we should.

3.1 Technology

The theme of technology occupies a central position in a number of Heidegger's later (i.e. post *c*. 1930) lectures and essays, notably his 1955 'Memorial Address' and the various papers collected in *The Question Concerning Technology*. In all of these works, the tone is distinctly critical: readers are left in no doubt that, in Heidegger's opinion, technology is in some sense a bad thing. Yet although he was no enthusiast for technological developments, Heidegger himself was certainly no neo-Luddite, and it would be wrong to portray his critique of technology as mere back-to-nature romanticism. His account is more complex, and more interesting, than that.

The first thing that must be noted, in considering that account, is that Heidegger uses 'technology' (*Technik*) as a term of art. To speak of 'technology' is usually to refer to a certain kind of artefact. So a television set or a fridge freezer or a carburettor – or even, perhaps, the famous Oncomouse – might be thought to count as technology, while a pebble or an acorn or an ordinary field mouse would not.

However, when Heidegger writes of technology he means to refer, not to any particular kind of entity, but to the way entities reveal themselves to us *as* entities. Technology, he writes, is a 'way of revealing' (1996: 318); it is the way the world tends to reveal itself to us late moderns.[1]

Whatever reveals itself technologically does so, Heidegger maintains, as 'standing reserve' (*Bestand*) (1996: 322), as, so to speak, 'standing by', ready to be put to use. To say that the present era is dominated by technology, in this sense, is therefore to say that nowadays entities tend to reveal themselves to us as mere resources, as things to be used and nothing more. Thus, to employ one of Heidegger's own examples, a modern-day forester might disclose himself as a 'human resource': whether he is aware of the fact or not, he is 'made subordinate to the orderability of cellulose', which is needed to make paper, which is needed, in turn, to produce newspapers and magazines, which are consumed by a gossip-hungry public, whose continued spending is required in order to buoy up the economy (cf. Heidegger 1996: 323; cf. Foltz 1995: 98*ff*). In a technological world, everything reveals itself as standing by, ready to be put to some use. Everything – even the forester – is there *for* something else.

Heidegger maintains that such a world may be associated with a distinctive mode of thinking, namely the kind that 'computes ever new, ever more promising and at the same time more economical possibilities' (1966: 46). And he adds that this 'calculative' mode of thinking goes hand in hand with a distinctive style of language, one that is nowadays epitomised by 'managerial' references to the provision of efficient or effective solutions, the rolling out of strategies, the meeting of targets and the production of neatly-packaged and quantifiable outcomes. The vocabulary will, I am sure, be familiar.

In these respects, then, Heidegger believes that the modern world is technological in character. Moreover, he believes that it is becoming increasingly so. For the technological way of revealing is, he claims, aggressive, like a cancer, in that it has the peculiar property of driving out other ways of revealing. So, to Heidegger's mind, the most unsettling feature of our modern age is not just that so many things are coming to reveal themselves as mere resources, but that more and more of them are coming to do so. The greatest danger, he insists, is that at some point *everything* might come to reveal itself technologically, that the world might become a bland and featureless expanse of standing reserve. This is the dystopia towards which, in his more pessimistic moments, he thinks we are heading.

Yet Heidegger concedes that things have not got quite this bad – at least not yet. Some aspects of life resist the encroachment of technology. Some marginal and 'inefficient' practices prove difficult to frame in terms of standing reserve (see further, Dreyfus 1993: 310–311). Backcountry trekking could provide one example: as

[1]Explaining what Heidegger means by this would take us too far off track. For present purposes it will suffice to note that by calling technology a way of revealing he does not mean to suggest that technology is merely a way we late moderns are prone to see the world. His main concern is not the anthropological one of describing different worldviews, but the transcendental one of investigating what Iain Thomson calls 'the conditions of the possibility of intelligibility' (2005: 54, n.15). See further, Thomson (2005: Chapter 1).

Thoreau long ago recognised, the exuberance and spontaneity of wild nature can provide a respite from the tidy and domesticated world of technology. However, Heidegger maintains that even our relations with nature are at risk of being technologised; indeed, to convey how far technology has encroached into modern life, he often chooses to focus on our attitudes to and treatment of the natural world. In the midst of technology, he writes, nature is becoming 'a gigantic gasoline station, an energy source for modern technology and industry' (1966: 50). The earth is coming to reveal itself as 'a coal mining district, the soil as a mineral deposit' (1996: 320). Living things are coming, more and more, to be exploited as 'stock' (1971: 112); even the air is being 'set upon to yield nitrogen' (1996: 320).

When nature reveals itself as a sink of resources it does so as so much standing reserve. Yet Heidegger maintains that the workings of technology are evident, not only in the consumption of 'natural resources', but also in the tendency to regard natural things as a source of what economists call non-consumptive use-values. Thus he bemoans the tourist industry (see, e.g. Safranski 1998: 401–3). After all, *en masse* (and often individually) tourists do not just consume vast quantities of food and drink and fossil fuels, they also see the sights, and as John Urry has pointed out, this activity amounts to a kind of consumption too: primarily, *visual* consumption (1995: 192). The eruption of Old Faithful in Yellowstone National Park, the Ganges from Varanasi at dawn, the view from the South Ridge of the Grand Canyon, the Great Pyramid of Cheops – visually consumed, fixed under what Urry refers to as 'the tourist gaze', these sites disclose themselves as little more than images to be ticked off an itinerary (cf. Tuan 1974: 95).

Were he alive today, Heidegger would no doubt observe that the workings of technology are evident, not merely in efforts to exploit the natural world, but also in the words and actions of those who seek to conserve it. Conservation, we are told, is about the proper *management* of natural *resources*. It is about assessing the economic value of natural *capital*, or of ecosystem *services* or *functions*. Champions of green business maintain, without blinking, that 'The environment is our core asset, the key component of product quality and an increasing priority for our consumers'.[2] Those who make such claims might sincerely wish to protect nature, of course. Nonetheless, theirs is a singularly technological approach to nature conservation (cf. Foltz 1995: 85).

3.2 Releasement

Heidegger maintains that the situation described above is our situation. We are in the midst of technology. What, then, does he recommend we *do*?

At first sight, it would seem that we cannot do much – or at least much that is useful – at all. Heidegger maintains that the way of revealing we call technology has

[2]Geoffrey Lipman, former President of the World Travel and Tourism Council, quoted in Neale (1998: 34).

been 'destined' for us by 'Being'. It is, he says, an episode in 'the history of Being', to which we are subject and from which we cannot free ourselves. 'Human activity,' he insists, 'can never counter this danger. Human achievement alone can never banish it' (1996: 339). It would therefore be foolish to condemn it as the work of the devil (1966: 53; 1996: 330).

Nevertheless, Heidegger does offer some hope. The dispiriting thought that technology has been destined for us can itself be transformed into a liberating realisation, a 'saving power'. We can realise that technology, though *destined* for us late moderns, need not be our *fate* (Heidegger 1996: 330). The ancient Greeks did not find themselves in a world of standing reserve; neither did our medieval forebears. The way of revealing we call technology is historically and culturally contingent. Accordingly, Heidegger maintains that even though it is impossible to overcome technology through sheer force of will, we can prepare the ground for the historical advent of a new way of revealing. We can maintain a receptive state of mind and in this way ensure that we will be able to notice and foster a new turn in the history of Being, should one come about.

One way we can do this is to change the way we comport ourselves towards the world – we can cultivate what Heidegger calls a releasement towards things (*Gelassenheit zu den Dingen*), which will enable us to 'let things be' (1966: 54). Much could be said about what Heidegger means by this (about, for instance, his intriguing idea that the 'being' of things is not provided gratis, as it were; but that things have to be *let be* by adopting a certain comportment towards them).[3] For now, however, it is enough to note that to be released towards a thing is, at least in part, to remain open to the various ways it could reveal itself to us – not just as a resource, but as an *objet d'art*, for instance, or as the focal point of a community. It is, in other words, to be receptive to the various meanings things have for us. Consider a glass of wine, for example. The thing might usually disclose itself to us as a resource and nothing more – as something to help one relax after a day's work, say. In normal circumstances, then, it reveals itself as being interchangeable with any other item that would serve the same function – a beer, for instance, or a cigarette. Yet Heidegger would insist that when even the most mundane of things is 'let be' it can come to intimate the presence of a number of dimensions of meaning. Thus, for one who is released towards it, the glass of merlot might intimate the grapes from which it was pressed, the rainwater and earth that nourished the vines on which those grapes grew; indeed, perhaps, the whole *terroir* of the land in which the vineyard lies. Moreover, it might intimate, not just all that, but a human world too: one in which glasses of wine are sipped nervously on first dates or raised in toasts at birthdays and weddings and retirement parties; one in which wine reveals itself, not merely as standing reserve, but as alive with a whole host of social and cultural meanings. Continuing speculations of this sort, Heidegger maintains that, when it is 'let be', even an ordinary everyday thing like a glass of wine can come to 'gather' a whole 'world' (see, e.g. 1971: 165–86). Hence, by letting a thing be, by releasing ourselves towards it, we can both release ourselves *from* the hegemony of technology

[3] On these issues, see James (2009: Chapter 4).

and release ourselves *to* the manifold dimensions of the thing itself, revealed now, not as standing reserve, but as the focal point of an entire field of meanings.

3.3 Conservation

At first sight, Heidegger's views might seem to have little bearing on what we would normally call conservation issues. When he bemoans the spread of technology, he is not expressing his fear that certain species might become extinct, say, or certain natural habitats destroyed. He would have conceded that these practical issues are important; however, he would no doubt have added that they are in the final analysis matters for politicians, scientists, economists and the like, rather than philosophers. Accordingly, Heidegger's dystopian vision of a world of standing reserve must be distinguished from the kind of doomsday scenario one sometimes finds portrayed in the environmental literature. A world in which technology reigns supreme could be in one in which endangered species have been conserved, in which the most harmful kinds of pollution have been curbed, in which various rich and valuable natural habitats have been saved. In other words, a certain kind of environmental *utopia* would, by the lights of Heidegger's account, constitute a technological *dystopia*.

Thus, although both Heidegger and the modern environmental thinker agree that the modern world is marked by loss, and although, moreover, they agree that the loss of specifically natural things is a particular cause for concern, they are referring to losses of different kinds. The environmental thinker wants to protect what might, broadly speaking, be called 'items' in the world (habitats, species, etc.), while Heidegger, for his part, worries that even those natural entities that remain in existence are coming to reveal themselves to us in only one way – namely, as standing reserve.

As we have seen, Heidegger maintains that in the face of this special kind of loss, we need to 'release' ourselves towards things and, by so doing, 'let them be', and this, I would like to suggest, constitutes a special kind of *conservation*. To conserve a thing in this sense is to conserve those of its dimensions that, in the rise and rise of technology, we are in danger of losing.[4]

This kind of conservation might appear, on the face of it, to be all about *value*. It might seem that to conserve nature in a Heideggerian manner is simply to retain our sense that nature has non-instrumental value, a value that is not merely a function of its usefulness. But this conclusion would be hasty. For one thing, talk of values can easily be framed in 'calculative' terms (cf. Heidegger 1977: 71). It is all too easy to think of moral and aesthetic value on the model of economic value, to imagine that it can be accumulated and lost like so many pounds and pennies. It is all too easy, then, for talk of values to contribute to the loss by which, if Heidegger is correct, the

[4]Cf. Bruce Foltz's decision to translate Heidegger's *Schonen* as 'conserving', a term sometimes translated as 'sparing' (1995: 161; cf. Heidegger 1996: 351).

modern world is marked.[5] Nonetheless, whether or not values-thinking lends itself to being framed in calculative or technological terms, it would seem that the sort of conservation we are here considering, the sort indicated by talk of releasement and 'letting be', is better thought of as a conservation of *meaning*. To be released towards the world is, as Heidegger puts it, to find oneself privy to 'the *meaning* [*Sinn*] which reigns in everything that is' (1966: 46; my emphasis and annotation).

The conservation of nature's meanings is, I suspect, best achieved through art and literature – not necessarily through the efforts of card-carrying artists and writers, but by anyone who succeeds in bringing to light the nature that lies outside the ever-expanding domain of standing reserve. Consider, by way of example, J.A. Baker's remarkable account of his efforts to follow peregrine falcons across the fields and marshlands of coastal East Anglia. Revealed technologically, those lands would appear in the light of some set of all-too-human instrumental concerns – as good for tourism, perhaps, or as useful flood defences. But *The peregrine* presents a world that is, in a certain sense, barely human at all. Not only are human beings generally absent, but as the book progresses Baker himself becomes more and more like the bird he is following. ('I sank into the skin and blood and bones of the hawk', he writes. 'The ground became a branch to my feet, the sun on my eyelids was heavy and warm' (2005: 144).) The results are strange and beautiful; the world evoked, a far cry from the relentless functionality and tedious, bullet-pointed efficiency of the world of technology. The same may be said of the writings of another nature writer, Richard Mabey. He writes of 'the layers of history bound up' in a woodland's 'grain and forkings and slow cycles of light and shade', of its 'aura of history', seeming to extend 'not just beyond what generations of humans have done to it, but beyond civilisation altogether' (2005: 11). This is the conservation of a meaning, a *natural* meaning, which to a less attentive commentator – and certainly to one obsessed with the valuation of natural services and functions – would have been lost. Or, as one more example, consider the following passage from (curiously enough) a third resident of East Anglia, namely, Mark Cocker:

> … the cattle can also create a vastly different mood. When the mist steals out of the dykes at twilight and winds gradually around them, it appears less that this shroud rises to engulf the cows than that the black beasts themselves are emerging from the Earth. The animals you presumed to know, with their comic threads of spittle and their comforting sweet straw smell, are now mysterious and new like giants resurrected in a primordial landscape. (2008: 142)

Cocker is attending to the meaning the cows have for us, but in so doing he is not ignoring the creatures themselves. He has not assumed the role of the bad romantic, for whom, as Mabey puts it, nature is 'little more than a vehicle for human feelings

---

[5]Heidegger himself rejected the practice of what he called 'values-thinking', maintaining that 'through the characterization of something as "a value" what is so valued is robbed of its worth' and (more hysterically) that 'thinking in values is the greatest blasphemy imaginable against Being' (1996: 251). Although there is no good reason to endorse Heidegger's view that values-thinking is inherently pernicious, there are, I believe, reasons for thinking that many of the ways in which nature matters to us – and, moreover, matters to us *morally* – are not best conceived in terms of value (see further, James 2009: Chapter 3).

and symbols' (2010: 188). Quite the opposite: Cocker's words testify to the same kind of selfless and non-anthropocentric attentiveness to nature that one finds, equally, in the works of Baker or Mabey (or Leopold or Dillard or Bachelard). Moreover, like the best works of these writers, his account doesn't just *support* the efforts of conservationists. It conserves nature - *directly*. True, it does not, in itself, stop these creatures from being wiped out of existence, nor does it directly prevent them from being harmed. Yet it does open our eyes to a constellation of meanings which, if Heidegger is correct, we are in danger of losing. Just as, in Cocker's nice image, the cows emerge from the Earth, so, through his well-chosen words, they come into focus, not just as providers of milk or as so many pounds' worth of stock, but as something older, stranger and more dignified.

3.4 Normativity

So, what does it mean to try to conserve nature? It can be to try to prevent certain vulnerable parts of the world – habitats, species, etc. – from being harmed, damaged or eradicated. However, it can also involve efforts to safeguard the various meanings nature has for us. Moreover, while the first sort of conservation requires the efforts of, for instance, scientists and environmentalists, the second is best achieved by writers and artists – by men and women like Baker, Cocker and Dillard.

If Heidegger is right – and I concede that I have merely suggested, rather than demonstrated, that he is[6] – then there is nowadays more opportunity than there once was for the second sort of conservation. With so many meanings being lost, there is more nowadays for the second sort of conservationist to do. Is there not simply more *opportunity* but more *need* as well? I have not so far engaged with normative matters. I have not maintained that nature's meanings ought to be conserved, still less that this 'ought' has *moral* force. I have merely argued that this is one way in which nature *can* be conserved. However, the subtext is hard to miss. The tone of my discussion has clearly implied that it is a good thing to conserve nature's meanings.

Yet this claim might seem dubious; for the meanings I have said we are in danger of losing, the meanings Baker, Mabey and Cocker manage so ably to conserve, are all what might be called *good* meanings, ones of which readers of a book like the present volume will approve. But of course not all nature's meanings are like this. To take one of countless examples: in the nineteenth century, American landscapes were often regarded as dull and barren because they seemed to be devoid of the historical and literary associations that served to enrich their more picturesque

[6] I do not have the space here to provide a thorough explanation of Heidegger's complex account of technology, still less to try to justify it (for an interesting attempt to do both, see Thomson 2005).

counterparts in the Old World (Saito 2004: 144). Such reactions are far less common nowadays. Surely this is no cause for regret? Surely this is one meaning that it is good to have lost?

No doubt it is good that we have largely relinquished the notion that, before the arrival of European settlers, America was an uninhabited wilderness. It is good, too, that we no longer tend to see the value of a natural landscape as being merely a function of its literary and historical associations. Indeed, in different times and cultures, nature has been invested with all sorts of meanings that reflect little more than ignorance and prejudice, and which are not worth conserving.[7] Yet to concede these points is not to undermine the case I have been making, for, although I have suggested that we ought to conserve nature's meanings, I have not denied that we need to exercise discrimination in doing so. Moreover, even if this need for discrimination is granted, the point remains that if Heidegger is right, then *our* danger, the danger we are facing here and now, is that nature may come to be stripped of all of its meanings, bar one. Faced with this possibility, it may be unwise for us to become *too* occupied with sifting 'good' meanings from 'bad' ones. Surely we ought, within reason, to conserve whatever meanings we can?

Again, a critic might ask what exactly we are to make of this *ought*. Why, precisely, are we obliged to conserve nature's various meanings? Even if, as Mabey (2010: 2) puts it, our vision of nature's meanings is becoming 'narrower', why should we regard this as a bad thing?

One could respond to such questions by appealing to the indirect benefits meaning-conservation has for nature. So, for instance, one could argue that although nature, in many of its various manifestations, ought to be conserved, people will only be inclined to conserve it if they see it as bearing certain meanings. Alternatively, one could appeal to the direct benefits meaning-conservation has for human beings. For example, one could argue that nature's meanings ought to be conserved because it is difficult – if not impossible – to live a truly worthwhile life unless one is disposed to see certain meanings in the natural world (cf. Holland, this volume). These are just two of the many ways one might try to justify the claim that we ought to conserve nature's meanings. Yet if that claim can be justified, another task remains: that of convincing people to release themselves *from* technology and *to* the semantic richness and depth of nature. Here, however, I suspect that rational persuasion will be of limited use. To be sure, such means of persuasion might have some effect on some people. But I suspect that rational arguments would have little effect on the man who concedes that nature is becoming drained of meaning and yet sees no reason to do anything about it. Better, perhaps, to lend him a copy of *The peregrine*.

---

[7]Think of the anthropomorphic tendency to see animal behaviour in the context of distinctively human norms, for instance, or the tendency to regard natural phenomena as possessing supernatural powers.

References

Baker, J. A. (2005). *The peregrine.* Introduction by R. Macfarlane. New York: New York Review Books.

Cocker, M. (2008). *Crow country.* London: Vintage.

Dreyfus, H. L. (1993). Heidegger on the connection between nihilism, art, technology, and politics. In C. B. Guignon (Ed.), *The Cambridge companion to Heidegger* (pp. 289–316). Cambridge: Cambridge University Press.

Foltz, B. (1995). *Inhabiting the earth: Heidegger, environmental ethics, and the metaphysics of nature.* New Jersey: Humanities Press.

Heidegger, M. (1966). *Discourse on thinking* (J. M. Anderson & E. H. Freund, Trans.). New York: Harper & Row.

Heidegger, M. (1971). *Poetry, language, thought* (A. Hofstadter, Trans.). New York: Harper & Row.

Heidegger, M. (1976). *What is called thinking?* (J. G. Gray, Trans.). New York: Harper Perennial.

Heidegger, M. (1977). *The question concerning technology and other essays* (W. Lovitt, Trans.). New York: Harper & Row.

Heidegger, M. (1996). *Basic writings* (D. F. Krell, Ed. & Trans.). Oxford: Blackwell.

James, S. P. (2009). *The presence of nature: A study in phenomenology and environmental philosophy.* Houndmills: Palgrave Macmillan.

Mabey, R. (2005). *Nature cure.* London: Chatto & Windus.

Mabey, R. (2010). *A brush with nature: 25 years of personal reflections on the natural world.* London: Random House.

Neale, G. (1998). *The green travel guide.* London: Earthscan Publications.

Safranski, R. (1998) *Martin Heidegger: Between good and evil* (E. Osers, Trans.). Cambridge, MA: Harvard University Press.

Saito, Y. (2004). Appreciating nature on its own terms. In A. Carlson & A. Berleant (Eds.), *The aesthetics of natural environments* (pp. 141–155). Toronto: Broadview.

Thomson, I. (2005). *Heidegger on ontotheology: Technology and the politics of education.* New York: Cambridge University Press.

Tuan, Y. (1974). *Topophilia: A study of environmental perception, attitudes, and values.* Englewood Cliffs: Prentice-Hall.

Urry, J. (1995). *Consuming places.* London: Routledge.

Chapter 4
Revaluing Body and Earth

Patrick Curry

> *...if intellect does not deserve the crown of crowns, only intellect is able to award it. And if intellect only ranks second in the hierarchy of virtues, intellect alone is able to proclaim that the first place must be given to instinct.*[1]

Abstract This paper argues for positively revaluing the body, the feminine, and the Earth as integrally linked aspects of a healthy *modus vivendi* that have been dangerously damaged by modernity. Merleau-Ponty's embodied phenomenology is presented as a promising way to avoid retaining its positionings of mind/spirit as comprising nothing but consciousness and sentience and the body as nothing but inert matter. For the feminine dimension of the project, I rely on Irigaray and Butler. The indispensability of animist agency is then maintained, with reference to Plumwood and Abram. From here, I argue for the centrality of places (as against space) and thence of the Earth, and an understanding of it that is not just ecological but fully ecocentric. Finally, the post-secular implications of the project are noted.

Keywords Embodiment • Place • Ecological phenomenology • Feminist philosophy • Animism

[1] Proust (1997: 25–26).

P. Curry (✉)
15 Hofland Road, London W14 0LN, England, UK
e-mail: pmcurry@gn.apc.org

E. Brady and P. Phemister (eds.), *Human-Environment Relations: Transformative Values in Theory and Practice*, DOI 10.1007/978-94-007-2825-7_4,

4.1 The Project of Revaluation

This paper suggests ways to begin revaluing the subjects in my title, as well as arguing that being internally linked, they cannot be revalued separately. I say 'suggests' for the subject is vast. Starting with the body, one could reasonably start by asking: are there any values which are not embodied? All human valuers are (or were at one time) embodied, although that does not mean, of course, that *what* they value need be restricted to the body. At the least, the embodiment of all human valuers surely suggests that the body is fully present in, and thus materially influences, not only the act of valuing but also, thereby, its outcomes. Giving recognition of this existential fact the emphasis it deserves but rarely gets, I shall refer not to 'embodiment' – as if someone else, or something else such as a 'mind', perhaps entirely different, is merely 'in' a body – but rather to 'bodiment' (Acampora 2006).[2]

What this 'body' is, then, cannot be taken for granted. In anticipation of the discussion of this question, I want briefly to note two important points. One is that since all human bodies are sex-gendered or sexuate,[3] sex in the fullest sense must also be considered. By implication, so should the work of those feminist philosophers and psychoanalysts who have made it their particular concern. The second point is that since all human bodies are fully embedded in particular concrete places, and all those places are either primarily or secondarily natural, nature too is integral to bodymind – something which invites attention to the work of ecological philosophers and some anthropologists.

Hence, there are really four subjects in play: bodiment, sexuality, place, and nature or the Earth. Without claiming it is the only or even best way to proceed, however, I will treat sex-gender as fundamentally an aspect of *bodiment*, and will move fairly smartly from place to *nature*.

Let us consider why such a project of revaluing matters. The answer will also help us understand how best to go about it. Its most important impetus and context is the increasingly unignorable contemporary ecocrisis, both 'inner' or psychosocial and 'outer' or environmental/ecological, whose dimensions in terms of climate change, crashing biodiversity and so on there is no need to review here. The accompanying destruction and degradation suggests the need to revalue what has been correspondingly downgraded by many discourses within Western philosophy in the last 350 years, although by no means without powerful older antecedents: the body, women and the feminine, and the natural world. Let me immediately add that by 'philosophy', I mean not only formalised thought, but also fundamental guiding metaphysical assumptions. By 'nature', I mean something other or more than an

---

[2]This relatively unified 'bodymind', or something very like it, can be found in much East Asian philosophy, but that is a thread which cannot be taken up here.

[3]Irigaray argues (and I agree) that an uncritical use of the sex/gender distinction reproduces an uncritical nature/culture distinction. See, e.g., Whitford (1991) and Irigaray (2004). But the relationship between 'biological' sex and 'cultural' gender remains difficult. It can be posed thus: the latter does not derive directly from the former; nor, however, is it related randomly or arbitarily…

inanimate and instrumentalized resource-base: all animate life, plus the biologically inanimate elements upon which it depends, the ensemble as well as home of which is the Earth. So, it is critically important that these items are understood as politically as well as substantively linked. Indeed, we could add another aspect of the same project: recovering the connections between theory, as one particular kind of practice, and other apparently more 'material' practices.

Corresponding to bodiment, the feminine, and Earthy nature are three interlinked dysfunctional or pathological discourses of theory and practice: idealism, spiritualism or culturalism (positioning the body as inferior etc.), androcentrism (positioning females and the feminine as inferior) and anthropocentrism (positioning nonhuman nature as inferior). All three modes are central to the exercise and protection of privilege – 'minds/men/humans are more valuable than everyone and everything else' (except possibly a God in whose image the human is supposedly made and who therefore licences such privilege) – and to the dynamics which have such toxic ecological consequences.[4] We could note, in addition, the determined effort by both modern science and capitalism to replace places (qualitative, plural, unique) with space, whose units are tendentially quantitative, single and interchangeable.[5]

It is thus far from simply a matter of compensation, of valuing the body, the feminine, place and the Earth more than they have been. What is at stake is precisely what these *are*. Here, we must admit the difficulty of recognizing, rethinking and revaluing embodied values, given the influences of Platonic, Christian and Cartesian or modernist values and concepts. Nor should the effects at stake be interpreted in purely metaphysical terms, insofar as the influence of imperial 'Western' cultural and material values have become global.

4.2 Pluralism

Still on the philosophical underlabouring that our task requires, it is worth noting that the value-laden dualisms I have already mentioned support, and are supported by, a dominant monism comprising both monotheism and the scientific truth that is its secular heir. In both cases, the effects begin with disenchantment and finally terminate in nihilism.[6] The dualisms operate by pitting spirit/mind/culture/humanity/male over and against materiality/body/nature/the non-human/female.[7] 'Culture' continues to do the work vis-à-vis 'nature' that 'spirit' used to do; although secularised, the underlying logic has not changed. Now although reversals of valuing, in which those

[4]For its exemplary critical attention to all of these modes and dynamics, see the work of the late Val Plumwood.

[5]See Casey (1997).

[6]As Weber, after Nietzsche, rightly perceived (see Curry 2007). Note, however, that monotheism retains a place for ultimate mystery that the latter's promise of ultimate mastery does not.

[7]See Plumwood (1993).

formerly or elsewhere subordinated or suppressed become ascendent and *vice-versa*, are possible, that does not change the underlying mode. It remains the case that there are two opposing and putatively exhaustive discourses each side of which aspires to become a monism: a single, universal and exhaustive truth.

Thus, as Eduardo Viveiros de Castro has noted, the true counter to monism, as well as to the vying dualisms that constitute and proceed from it, is pluralism. ('True' in the double sense of both conceptually and effectively.) Here, unfortunately, I am obliged to abbreviate a complex subject by invoking other and/or previous work: specifically, Viveiros de Castro's 'perspectivism', Bruno Latour's (1993) 'relationism', what Paul Feyerabend (1987) and Barbara Herrnstein Smith (1988, 1997) call 'relativism', and what I have termed 'relational pluralism' (Curry 2003, 2008).[8]

As those terms imply, 'relations are fundamentals, relata are abstractions' (Weber 2005: 215).[9] I do not see this as an obstacle to retaining 'value' (and especially 'intrinsic value', or some equivalent) as a particular and important kind of relational meaning. Substantively, the debate between 'objectivist' and 'subjectivist' accounts of value is often misleading. Valuing is an act and therefore involves a valuer (who need not be human, of course), but it does not follow from this that value does not *also* inhere in what is valued. Nor does it follow that value is entirely subjective or voluntaristic, as if no world or others existed with their own formations, agencies and agenda. Indeed, to grant those terms of the debate is to accede to just what is being contested: dualism in the service of monism. Thus, items are intrinsically valuable when they are valued for their own sake without reference to their usefulness in realising some other end. Furthermore, strategically, it is vitally important that we retain the concept of intrinsic value. Abandoning it would leave the natural world – including that part or dimension of human beings which falls into that purview – even more defenceless than it already is against being instrumentalized, appropriated and exploited. Any approach that helps limit the dynamically destructive as well as occasionally creative human tendency and ability to meddle is not to be lightly surrendered.[10]

Among the radical consequences of relational pluralism is a highly desirable (not to say long overdue) shift from epistemology, and *a fortiori* methodology, to ontology: or rather, since perspectives are always plural and pluralism is perspectival, a shift to ontolog*ies*, that is, from a putatively single knowing of one ever-retreating notional world to many particular and participatory worlds and ways of being in and of them.[11]

A corollary of this way of subverting dualisms is to realise that distinctions between 'the material' and 'the cultural' are not foundational, through recognising the potential, and on occasion actual, subjectivity and agency of the so-called

[8]There is also a degree of pluralism at work in environmental pragmatism; for a brief recent discussion, see Curry (2011: Chap. 10).

[9]It is significant that this statement occurs as part of a lucid summary of William James's philosophy.

[10]See Kane (1998: 50).

[11]See Viveiros de Castro (2004), and note the resonance with later Wittgenstein. (Methodology itself, as Mary Midgley once observed, tends to morph into methodolatry.)

material (and not attributing these aspects to it, as if it were entirely within our remit to graciously grant or withhold them). To quote Val Plumwood, 'materiality is already full of form, spirit, story, agency, and glory' (2002: 226). The same point applies to the embodied and material dimension of culture's hitherto most idealistically privileged domains, e.g., higher mathematics, advanced theorizing of all kinds, and spiritual experience and discourse.[12]

It is well to note that a double move is required, for to redefine the material as also cultural must be accompanied by a move to redefine the cultural as also material. Only thus can the twin dangers be avoided of tendentially complete naturalization on the one hand, e.g. evolutionary psychology, and tendentially complete culturalization on the other, e.g. some forms of philosophy, cultural studies, and spirituality (including much of the 'New Age' kind). Both positions are unwarranted essentialisms. Both, to borrow a remark of Bateson and Bateson (1987: 51), are modern superstitions, and both support a ruinous split that must be contested and ultimately healed.

Lest the preceding remark arouse suspicions of naïve anti-Cartesianism, let me add that contingent local distinctions between spiritual or mental and material, between natural and cultural or, for that matter, between real and unreal are not the problem, any more than are either rationality or spirituality *per se*. It is their forced and enforced conversion into an ideology and programme (rational*ism*, spiritual*ism*, etc.) which is pathological.[13]

That point has two aspects which require attention. One is that it is perfectly possible, and probably always has been, for human beings to distinguish between the mental or spiritual (hence 'cultural') on the one hand and the physical or material (hence 'natural') on the other, even if the distinctions were much more fluid and contextual than they have since, and in some places, become. The other is that, as already noted, they have become incorporated into an ideology – or rather, an ideological lineage, the most recent instance of which can reasonably be called 'modernism' – which urgently requires questioning and indeed subverting. Together, these two aspects raise a critical question for any such project: how do we give due weight to the first, human point without accepting the proclamations of its absolutist twin? More generally, in this project of reconceiving and revaluing, there is a double danger to avoid: succumbing to ingrained culturalist/spiritualist misapprehensions of the material and natural, on the one hand, and to naturalist/materialist misapprehensions of the spiritual and cultural, on the other.

4.3 Merleau-Ponty's Phenomenology

With that warning in mind, let us return to bodiment, the putative instantiation of nature and matter, and its relationship with its distinguishable but inseparable twin, and the putative bearer of culture and spirit or mind, or what we could call

[12] See, e.g. Csordas (1994); Lakoff and Johnson (1999).

[13] See Kontos (1994).

'mindment'. Faced with the question of how to think intelligently about this subject, the work of Maurice Merleau-Ponty is decisive, so it is to this that I now turn.[14]

To oversimplify a complex and subtle philosophy, Merleau-Ponty argued that while perceiving subject or 'mind' and perceived object or 'body' (whether one's own or others') are not one and the same, neither are they completely separate. Rather these two poles of 'body-subjects' are connected by a constitutive divergence or gap (*écart*) which is chiasmically criss-crossed, intersected, entwined, and by virtue of which both perceiving *subject* and perceived *object*, that is, both *self* and *other* are neither entirely identical nor entirely different. This means that, although we can meaningfully speak of one pole or the other, we cannot do so without reference to its complement, nor can we do so in a way that takes either pole to be self-identical. Instead, together the two poles comprise a decentred duality which is not, however, a dualism.[15] Merleau-Ponty's term for this dynamic, as he developed it in his posthumously-published work, was *Flesh*. Moreover, Flesh is ontological, insofar as the subject-object distinction and the attendant concern with representation that characterise epistemology arise from it, and are therefore secondary to it.[16]

The dynamic of Flesh, while it prevents both pure subjectivity and pure objectivity, is also what makes these possible at all. Conversely, 'what enables us to center our existence is what also prevents us from centering it completely…' (Merleau-Ponty 1962: 85). However, the key point, in the present context, is that there is a minded dimension to the body, and a bodied dimension to the mind, which in both cases is inalienable and inherent. A more promising way out of the schizogenic modernist impasse, a way that not only recognizes the distinctiveness of each but also their fundamental intertwinedness, can hardly be imagined.[17]

For instance, it encourages us to give due weight to the rootedness of even our most abstract thoughts and theories in bodily perceptual processes: inside (inclusion)/outside (exclusion), up (higher)/down (lower), centre/periphery, attraction (sympathy)/repulsion (antipathy), symmetry/asymmetry, declivity/aclivity, paths, blockages, links, scales and cycles among others.[18] Yet crucially, Merleau-Ponty's approach resists the scientistic reduction of 'mind' *to* 'body', a reduction that is familiar to us as the attempt, starting from a dual*ism*, to enforce a materialist monism. Thomas Csordas makes the point, in similarly non-naturalistic and counter-hegemonic vein, that 'our bodies are not originally objects to us. They are instead the ground of perceptual processes that end in objectification' (1994: 7).

This perspective thus encourages recognition of the extent and ways in which the body cannot be considered apart from, or *as* apart from, the discursive. It is not a

[14]See Curry (2010a) for a more extended discussion in which I also supplement Merleau-Ponty with, and relate his work to, that of Paul Ricoeur on metaphor.

[15]It is certainly not the classic Cartesian dualism of body as extended but inert versus mind, spirit or self as sentient but unlocated. For good discussions, see Reynolds (2004) and Hass (2008).

[16]See Olkowski (2006: 13).

[17]Although see too Latour (1993).

[18]Some of which I have taken from Johnson (1987: 206).

question (as I believe the work of Michel Foucault implies) of the relatively passive body being molded or inscribed by the latter, even if that involves auto-inscription. The body for Merleau-Ponty is both active and alive *and* fully cultural.[19]

4.4 Feminist Perspectives

One of the principal concerns of feminist philosophy is bodiment, which it treats both positively (as an important subject in its own right) and negatively (reacting against the tradition of its dismissive treatment, and/or interested appropriation, in philosophy dominated by men). Hence, it is not surprising that Merleau-Ponty's work has found a receptive but critical audience amongst feminist philosophers. To evaluate all the claims involved is beyond the scope of this paper, so I will confine the discussion to some important ideas of Luce Irigaray – formed partly in response to Merleau-Ponty – as refracted in a recent paper by Judith Butler.[20]

There is in feminist philosophy a longstanding debate over how to recognize and perhaps reconcile two potentially conflicting perspectives: first, the distinctiveness of female and male bodies and the profound psychosocial consequences that follow therefrom; and second, the valuable sceptical critique, postmodern and earlier, of essentialism. This dilemma was apparent in earlier debates on how to interpret Luce Irigaray's philosophy, perhaps especially in the argument (whether directly or indirectly) between the apparently biomystical essentialism of Irigaray herself and Judith Butler's (1994) uncompromising discursive and political anti-essentialism.[21]

Compressing the issue greatly, Irigaray suspects Merleau-Ponty's chiasm of concealing significant sexuate differences, and thereby falsely universalising a masculine mode and an anonymity over what is really the maternal-feminine. The result privileges commonality over difference in a way that borders on solipcism.[22] Butler, however – and in my view, correctly – understands chiasmic intertwining as constituting both masculine and feminine, each of which is non-identical with itself thanks to its integral dependence on the other. By the same token, the bodied and chiasmic self is integrally implicated in alterity (others whom, and worlds which, it does not control); and the maternal body is equally constituted by such relations. To be intertwined in such a way rules out not only beings' absolute autonomy, and therefore solipcism, but also their reducibility one to the other.

---

[19] As for how this link can best be understood, although there is no room to develop the idea here, I agree with Csordas (1994: 16) again that 'the critical meeting ground between textuality and embodiment' is metaphor (see my 2010). I also suspect that Michael Polanyi's 'tacit knowledge' and consequent 'post-critical' philosophy might provide fruitful insights.

[20] Irigaray (1987) and Butler (2006); see also Butler (1989). For a convincing refutation of some feminist critiques of Merleau-Ponty, see Stoller (2000).

[21] See Butler (1994) and the other essays in Burke et al. (1994), as well as Stone (2006).

[22] Cf. Levinas (1977: 55). For a good critique of Levinas's criticism of Merleau-Ponty, see Hass (2008: 112–122, 132–133).

To conclude, bodiment, for and as humans, is fully sex-gendered or sexuate. But there are some bodied human commonalities that are not sexuate; so perhaps we could say, adapting a remark by Derrida about politics, that everything about us is sexuate, but it is not *only* that. However, no account of human bodiment can claim to be comprehensive without engaging non-reductively with sexuality is this sense. Furthermore, such engagement renders undeniable the asymmetry of male and female natures, experiences and worlds, and the profound psychological, psycho-analytical, social, political and philosophical consequences of that fact. In a nutshell, both men and women are embodied but only the latter give birth to embodied life; so we might say that the female makes both itself and the male possible, although that is not what it, and/or she, is 'for'. Relations between them are also reciprocal but asymmetrical, just as both proceed developmentally in terms of bodiment – but not in the same terms. Finally, it is possible, and fruitful, to understand relations between the two sexes/genders as themselves chiasmically intertwined: neither wholly unrelated nor reducible one to the other, neither hyper-separated nor an ultimate identity, and neither one self-identical.

4.5 Agency and Animism

As his last work makes clear, Merleau-Ponty's ontological and dynamic Flesh is both of bodies and of the world. Furthermore, what he also described (1968: 203) as 'wild Being' (*l'être sauvage*) is ontologically animate. These points put the body and bodiment, so conceived, squarely in the ecological domain.

Consideration of body-as-world, and world-as-body, connects directly with the pluralism and perspectivism with which we started, insofar as the body, as 'an assemblage of affects or ways of being that constitute a habitus', is precisely the seat of perspectives, and as many as there are bodies; hence the pluralism. Furthermore, since 'a perspective is not a representation', this Merleau-Pontian meta-perspective, together with those of ecofeminism and the new animism, subverts the modernist obsession with, and fetishization of, epistemology, and its claim to be able to exhaustively represent 'the' world.[23] On the contrary, and counter-hegemonically, these approaches offer a recovery from 'ontological poverty… a poverty characteristic of modern thought' (Viveiros de Castro 2004: 475, 474, 482–483).

The epistemological claim to represent anything accurately and exhaustively, even in principle, presupposes the ability of humans (and in practice, only a very few of them) to be able to do so. Such a programme is embedded in a monism reiterating, however unconsciously, its monotheistic provenance; in modernist terms, 'its ideal is the system from which all and everything follows' (Horkheimer and Adorno 1994: 7), a role which theology used to fill. Insofar as both theistic and modernist monism assume a fundamental difference between spirit and matter, the

[23] Note that vulgar relativism offers no real challenge, insofar as it simply denies that truthful or accurate representation is possible; the debate thus remains on the debilitating ground of epistemology.

only outcome on offer is either a thorough-going idealist one, in which human subjectivity somehow transcends materiality (including, of course, the body) and achieves full disembodied autonomy, or a thorough-going materialist one, in which objectivity, i.e. objectness, supposedly excludes any vestiges even of subjectivity or agency, including that of the materialists themselves! Both idealism and materialism are on offer today in various forms, but both suffer from what Herrnstein Smith has anatomised as 'intellectual/political totalitarianism (the effort to identify the presumptively universally compelling Truth and Way and to compel it universally)…' (1988: 179). As a result, their effects are almost equally pathological, and specifically anti-ecological.[24]

In this context, a significant struggle is to reconnect with a post-technoscientific nature which is once again – or better still, remains – a subject as well as an object. Two vital concepts are required for this effort. In explicating them, I shall call upon the work of two contemporary ecological philosophers, Val Plumwood and David Abram respectively.[25] *Agency*, as basically cognate with subjectivity but, crucially, a bodied subjectivity, is non-anthropocentric. I have already quoted Plumwood's recognition and revaluation of the material (and relatedly, the natural and the feminine) as fully agentic. In an explicit attempt to bypass definitions of agency which, invoking our classical Greek heritage of valorising anthropocentric as well as androcentric 'reason', emphasize cognition in order to extend human exceptionalism, Plumwood (2006a: 124) defines agency simply but powerfully as 'active intentionality'. This radically non-Cartesian and non-modernist approach keeps the door open to the agency and indeed spirituality (embodied and relational) of all and any others: other animals, for example, but also, recalling Merleau-Ponty's Flesh as the wild Being of/as the world, places – a point to which we shall return. Such agency cannot be confined to subjects-which-are-(apparently)-not-also-objects, minds-which-are-(apparently)-not-also-bodies, or spirit-which-is-(apparently)-not-also-matter. Hence, agency is equally characteristic of what we take to be objects, bodies and matter; and is not, of course, restricted only to human bodies.[26]

The other closely-related concept is *animism*. Rescued from 'the enormous condescension of posterity'[27] – in this case, the teleological imperialism of early philosophical anthropologists – the new animism, breaking with the latterly modern aberration of assigning mind or soul to humans alone, is metaphysically egalitarian or democratic. It is also, crucially, non-anthropocentric or ecocentric.[28] Here the

[24]On this point, Max Horkheimer and Theodor Adorno's analysis in *The Dialectic of Enlightenment* remains hard to improve upon, even if it requires supplementing by, say, Latour (1993).

[25]'Environment' and 'environmental' are unhelpful terms, reducing as they do – in a manner closely parallel to 'embodiment' – a non-natural being merely surrounded by non-human nature. 'Ecological', while not without its problems too, is preferable.

[26]In terms of Latour's sometime 'Actor Network Theory', agency is a property of networks rather than *any* particular item as such. And such networks are fully as material as they are ideational or imaginal.

[27]E.P. Thompson's still-resonant phrase.

[28]See Harvey (2006). Also see Ingold (2006) and Hornburg (2006). On ecocentrism see Curry (2011).

ecological phenomenology of David Abram comes to mind, since his influential book *The Spell of the Sensuous* is based on the fact, obvious in hindsight but momentous at the time of writing, that the Merleau-Pontian phenomenological body is metaphorically and literally inseparable from the natural world of which it is a part, and without which it cannot exist. 'Intelligence is no longer ours alone but is a property of the earth; we are in it, of it, immersed in its depths…' (1996: 242). Hence, we arrive at the world as more-than-human: an important reminder that an ecocentric ambit includes, but radically qualifies, a human one, and that our sanity, mental and social as well as ecological, requires just such a context.[29]

Animism is a good term, then, for the appropriate mode of relationship to, and participation in, such a world. Notice, however, that the upshot is *not* a programme of systematic or universal animism, whether in theory or in practice. That would amount to yet another monism, lethal to the phenomenon itself *as* something or someone alive and agentic – not just notionally but experientially, that is, 'phenomenologically'.

On this point, Irving Hallowell's account of his exchange with an Ojibwe elder transcends amusing anecdote to become a salutary warning. Although well-known in some anthropological circles, it bears repeating:

> In the 1930s Irving Hallowell asked an unnamed old man among the Ojibwe of the Beren's River in Manitoba, 'Are *all* the stones we see here about us alive?' Hallowell continues, 'He reflected a long while and then replied, "No! But some are."'[30]

That answer (including its air of polite incredulity) is as correct as the unhelpful way the question was framed permits. In other words, it is not that all stones, or anything else, are necessarily *or* universally alive, but rather that *anything can* be experienced as alive and, by the same token, be a subject with whom one is, as a subject oneself, in a relationship.[31] As Graham Harvey writes, 'Animists are people who recognise that the world is full of persons, only some of whom are human, and that life is always lived in relationship with others'. Furthermore, they recognize 'the aliveness of all who live *as*, as well as in, particular lands or places' (Harvey 2006: xi, 19 [my emphasis]).

4.6 Place

This last point suggests another shift of emphasis. Animist relations, as we have seen, can be with virtually anything, including places. However, in this construal, what matters is not so much places in terms of technically complex systems

---

[29] Here, as so often throughout this sort of discussion, Gregory Bateson comes to mind. More recently, Abram (2010: 108–109) has affirmed the important point that redefining intelligence as bodied falls far short, and courts not just inconsistency but an ugly speciesism, if it fails to recognise that bodied intelligence is not limited to human bodies. In short, it must be ecocentric.

[30] Taken from Harvey (2006: 33).

[31] The result is sometimes an experience of enchantment, with certain implications for re-enchantment. I intend to explore this in depth in a future book.

(i.e. congeries of smaller-but-more-fundamental items, themselves inanimate) – not even places as ecosystems – but as *beings* themselves; and as such, both, albeit chiasmically, bodied and minded.

To put it another way, there is a profound elective affinity between body and place which this perspective reveals. Bodies are not only always, necessarily, *in* places, they may also be considered *as* relatively autonomous, e.g. mobile, places themselves. Conversely, places are, in an important sense, bodies too: not the universal, abstract, quantitative, causal stuff of 'space', but particular, sensuous, qualitative, participatory beings. Moreover, insofar as beings are constituted by relations, it follows that relationships with any place, and thereby knowledge of it, are only possible by virtue of participating in them as another body-subject. As Edward S. Casey puts it, 'to be is to be in place – *bodily*' (1997: 340).[32]

Knowledge of places is therefore necessarily perspectival. This is by no means a poor second-best option if complete knowledge from a view from nowhere (or what amounts to the same thing, from everywhere) is a fantasy. As Merleau-Ponty noted, spotting the modern emperor's nakedness: 'All my knowledge of the world, even my scientific knowledge, is gained from my own particular point of view' – both bodied and placed – 'or from some experience of the world without which the symbols of science would be meaningless' (2002: ix). Rejecting this recognition, the socio-historical process of converting place into space – as part of the overarching programme to replace ambiguity and partiality with putative certainty and security – is one of the lynchpins of modernity as a dominant and hegemonic formation, albeit one that is uneven, incomplete and unstable, with all the destructive consequences for the health of our body-minds, fellow-beings and places that are increasingly evident.[33]

4.7 Post-secularism

There remains one final aspect of embodied relational pluralism to emphasize, namely its *post-secularism*. This results from the coincidence of two considerations. First, a determination to be more consistent than those postmodernists, such as Richard Rorty, who have retained and protected the modernist (including scientistic) commitment – itself scientifically ungroundable, without fatal question-begging – to secularism and/or atheism. Second, the considerable amount of ethnographic evidence of various kinds which is relevant to the nature and experience of the liminal and, thence, of the sacred.[34]

Since body-minds of all kinds are constituted relationally and contextually, it follows that they are not only formed by chiasmic *écarts*, but are also themselves

---

[32] See also Casey (2009). Casey's work is itself influenced by Merleau-Ponty, among others.

[33] As well as Casey (1997), see Toulmin (1990) for an excellent account of this and related historical processes.

[34] This work was, of course, begun by Arnold van Gennep and Victor Turner.

such places. They are gaps or thresholds (*limns*) criss-crossed by, so to speak, interacting lines of relation in motion.[35] That is, they unavoidably entail boundaries which both separate *and* conjoin them to other bodies and places, and which make their existence both possible and vulnerable. In this process, no one party is completely in control; indeed, if one party were in control then there would not be, by definition, a relationship.

The 'lines', 'motion', 'boundaries' here are metaphorical, but not thereby 'merely' so, there being no non-metaphorical alternatives.[36] One could well say that what they refer to is itself metaphor in its actual operations, although of course, one could always legitimately argue for *different* metaphors for metaphor.

Now it is hardly surprising, given the vital importance of such dynamics for all life, that, to quote the neo-platonist Porphyry, 'Every threshold is sacred'. The point is amply confirmed by much anthropology and the humanities, while obscured, incompletely but far too successfully, by the modernist ideology of disenchantment. Notwithstanding the latter's emphasis on creation/production and commodity, the natural economy of the body-mind-place remains one of transformation, of exchange and the gift. Maintaining its integrity while negotiating necessary but perilous exchanges demands not only knowledge, but also skill, courtesy and humour. In short, it demands wisdom.[37]

The creation, learning and transmission of such wisdom is integral to ritual, just as ritual is integral to both individual and collective culture, and perhaps especially so in relation to the effectively sacred. However, let me remind the reader that 'culture', as understood here, cannot consistently be construed as hyperseparated[38] from 'nature' (although a unity of the two is an equally misleading idea). Parallel with body-minds, there are equally chiasmic 'nature-cultures' (Latour 1993).

Both strategically and substantively, then, the understanding advocated here is post-secular.[39] It recognizes and admits the ultimate unmasterable mystery 'that there is anything', as well as the existential mysteries of birth, sex and death. But this sacred or divine is not transcendental. Rather, it is immanent in – or rather, as – everything. That is to say, everything embodied is immanent, not in a logically necessary, e.g. Spinozan sense, but instead as an unrestricted potentiality, an openness to encountering living agency – and to its numinous intensification, the sacred or divine – in any particular relation and situation. (This returns us to what Plumwood calls 'a materialist spirituality of place' (2002: Chap. 10)).

Parenthetically, there is a politics implicit in the revaluing we have discussed here. It is not a politics based upon either the hyperseparation of self-mind-male-human and other-body-female-nature. Nor is it a politics based on their absolute unity, as

[35] For a fascinating recent discussion of the importance of 'motility' see Holbraad (2007), as well as Ingold (2006).

[36] See Ricoeur (2003).

[37] See Curry (2010b).

[38] Val Plumwood's (1993) useful term.

[39] See Curry (2007).

some Deep Ecologists propound. Bodied and (em)bedded more-than-human sentience, or Flesh, as should be obvious by now, rules out both these options. Rather, the politics implicit in the revaluing here is one of 'solidarity, the most fundamental of political relationships' (Plumwood 2006b: 70), which finds relation, or recognises kinship, notwithstanding profound differences. I find this obscurely but unmistakably encouraging.

References

Abram, D. (1996). *The spell of the sensuous: Perception and language in a more-than-human world*. New York: Vintage Books.

Abram, D. (2010). *Becoming animal: An earthly cosmology*. New York: Pantheon Books.

Acampora, R. R. (2006). *Corporal compassion: Animal ethics and philosophy of body*. Pittsburgh: University of Pittsburgh Press.

Bateson, G., & Bateson, M. C. (1987). *Angels fear. An investigation into the nature and meaning of the sacred*. London: Rider.

Burke, C., Schor, N., & Whitford, M. (Eds.). (1994). *Engaging with Irigaray. Feminist philosophy and modern European thought* (pp. 141–173). New York: Columbia University Press.

Butler, J. (1989). Sexual ideology and phenomenological description: A feminist critique of Merleau-Ponty's phenomenology of perception. In J. Allen & I. M. Young (Eds.), *The thinking muse: Feminism and modern French philosophy* (pp. 85–100). Indianapolis: Indiana University Press.

Butler, J. (1994). Bodies that matter. In C. Burke, N. Schor, & M. Whitford (Eds.), *Engaging with Irigaray. Feminist philosophy and modern European thought* (pp. 141–173). New York: Columbia University Press.

Butler, J. (2006). Sexual difference as a question of ethics: Alterities of the flesh in Irigaray and Merleau-Ponty. In D. Olkowski & G. Weiss (Eds.), *Feminist interpretations of Maurice Merleau-Ponty* (pp. 107–126). University Park: The Pennsylvania State University Press.

Casey, E. S. (1997). *The fate of place: A philosophical history*. Berkeley: University of California Press.

Casey, E. S. (2009). *Getting back into place: Toward a renewed understanding of the place-world* (2nd ed.). Bloomington: Indiana University Press.

Csordas, T. J. (1994). Introduction. In T. J. Csordas (Ed.), *Embodiment and experience: The existential ground of culture and self*, Cambridge: Cambridge University Press.

Curry, P. (2003). Re-thinking nature: Towards an eco-pluralism. *Environmental Values, 12*(3), 337–360.

Curry, P. (2007). Post-secular nature: Principles and politics. *Worldviews: Environment, Culture, Religion, 11*(3), 284–304.

Curry, P. (2008). Nature post-nature. *New Formations, 26*, 51–64.

Curry, P. (2010a). Embodiment, alterity and agency: Negotiating antinomies in divination. In P. Curry (Ed.), *Divination: Perspectives for a new millennium* (pp. 85–118). Farnham: Ashgate.

Curry, P. (2010b). Grizzly man and the spiritual life. *Journal for the Study of Religion, Culture and Nature, 4*(3), 195–209.

Curry, P. (2011). *Ecological ethics: An introduction* (2nd ed.). Cambridge: Polity Press.

Feyerabend, P. K. (1987). *Farewell to reason*. London: Verso.

Harvey, G. (2006). *Animism*. London/New York: C. Hurst & Co/Columbia University Press.

Hass, L. (2008). *Merleau-Ponty's philosophy*. Bloomington: Indiana University Press.

Holbraad, M. (2007). The power of powder: Multiplicity and motion in the divinatory cosmology of Cuban Ifá (or *Mana*, Again). In A. Henare, M. Holbraad, & S. Wastell (Eds.), *Thinking through things: Theorising artefacts ethnographically* (pp. 189–225). London: Routledge.

Horkheimer M, & Adorno, T.W. (1994 [1944]). *The Dialectic of enlightenment.* New York: Continuum.

Hornburg, A. (2006). Animism, fetishims, and objectivism as strategies for knowing (or not knowing) the world. *Ethnos, 71*(1), 21–32.

Ingold, T. (2006). Rethinking the animate, re-animating thought. *Ethnos, 71*(1), 9–20.

Irigaray, L. (1987). The politics of sexual difference (S. Hand, Trans.). In T. Moi (Ed.), *French feminist thought.* London: Blackwell.

Irigaray, L. (2004). *Key writings.* London: Continuum.

Johnson, M. (1987). *The body in the mind: The bodily basis of meaning, imagination, and reason.* Chicago: University of Chicago Press.

Kane, S. (1998). *Wisdom of the mythtellers* (2nd ed.). Peterborough: Broadview Press.

Kontos, A. (1994). The world disenchanted, and the return of Gods and Demons. In A. Horowitz & T. Maley (Eds.), *The barbarism of reason: Max Weber and the twilight of enlightenment* (pp. 223–247). Toronto: University of Toronto Press.

Lakoff, G., & Johnson, M. (1999). *Philosophy in the flesh: The embodied mind and its challenge to Western thought.* New York: Basic Books.

Latour, B. (1993). *We have never been modern.* Hemel Hempstead: Harvester Wheatsheaf.

Levinas, E. (1977). *Collected philosophical papers.* The Hague: Martinus Nijhoff.

Merleau-Ponty, M. (1968). *The visible and the invisible.* Evanston: Northwestern University Press.

Merleau-Ponty, M. (2002 [1962]). *The phenomenology of perception.* London: Routledge.

Olkowski, D. (2006). Introduction: The situated subject. In D. Olkowski & G. Weiss (Eds.), *Feminist interpretations of Maurice Merleau-Ponty* (pp. 1–24). University Park: The Pennsylvania State University Press.

Plumwood, V. (1993). *Feminism and the mastery of nature.* London: Routledge.

Plumwood, V. (2002). *Environmental culture: The ecological crisis of reason.* London: Routledge.

Plumwood, V. (2006a). The concept of a cultural landscape: Nature, culture and agency in the land. *Ethics and the Environment, 11*(2), 115–150.

Plumwood, V. (2006b). Feminism. In A. Dobson and R. Eckersley (Eds.), *Political theory and the ecological challenge* (pp. 51–74). Cambridge: Cambridge University Press.

Proust, M. (1997). *Marcel Proust on art and literature 1896–1919* (Sylvia Townsend Warner, Trans.; 2nd ed.). New York: Carroll and Graf.

Reynolds, J. (2004). *Merleau-Ponty and Derrida: Intertwining embodiment and alterity.* Athens: Ohio University Press.

Ricoeur, P. (2003). *The rule of metaphor. The creation of meaning in language* (R. Czerny, Trans.). London: Routledge.

Smith, B. H. (1988). *Contingencies of value: Alternative perspectives for critical theory.* Cambridge: Harvard University Press.

Smith, B. H. (1997). *Belief and resistance: Dynamics of contemporary intellectual controversy.* Cambridge: Harvard University Press.

Stoller, S. (2000). Reflections on feminist Merleau-Ponty skepticism. *Hypatia, 15*(1), 175–182.

Stone, A. (2006). *Luce Irigaray and the philosophy of sexual difference.* Cambridge: Cambridge University Press.

Toulmin, S. (1990). *Cosmopolis: The hidden agenda of modernity.* Chicago: University of Chicago Press.

Viveiros de Castro, E. (2004). Exchanging perspectives. *Common Knowledge, 10*(3), 463–484.

Weber, M. (2005). James's non-rationality and its religious extremum in the light of the concept of pure experience. In J. Carrette (Ed.), *William James and the varieties of religious experience* (pp. 203–220). London: Routledge.

Whitford, M. (1991). *Luce Irigaray: Philosophy in the feminine.* London: Routledge.

Chapter 5
Hölderlin and Human-Nature Relations

Alison Stone

Abstract In this chapter I outline a dilemma. If we resist human/nature dualism by thinking of human beings as embodied and environmentally situated – as natural – then this seems to imply that, as human products, modern environmentally damaging cultures are after all natural. However, claiming that these cultures are alienated from nature seems to return us to belief in human separateness. I explore a solution to this dilemma found in the work of Friedrich Hölderlin. He suggests that nature has divided itself into human subjects and the particular, finite, natural objects of which we are conscious and from which we feel separate. Thus both humanity and natural environments are aspects of nature in an expanded sense, as a self-dividing whole.

For Hölderlin, then, human estrangement from nature – and by extension contemporary environmental crisis – is a product of nature itself. This has the seemingly unhelpful implication that we human beings neither can nor should attempt to prevent this crisis. However, I argue that this quietism is less unhelpful than it might seem. It is motivated by anti-anthropocentrism; it anticipates Heidegger's scepticism about seeking a technological fix; and it allows for a novel justification of environmentalist practices as ways of preparing for possible change in nature's way of being.

Keywords Hölderlin • Nature • Poetry • Anthropocentrism • Technology

In this chapter I explore a dilemma that arises when we attempt to reconceive human beings as embedded within, rather than separated from, natural environments. Suppose that we describe one reason why modern societies are in environmental crisis by saying that their members are cut off, alienated, or separated from nature, rather than living and experiencing themselves as embedded within it. For this separation to have become possible, we might think, human beings must have already

A. Stone (✉)
County South, Lancaster University, Lancaster LA1 4YL, England, UK
e-mail: a.stone@lancaster.ac.uk

E. Brady and P. Phemister (eds.), *Human-Environment Relations: Transformative Values in Theory and Practice*, DOI 10.1007/978-94-007-2825-7_5,

been in a state of at least potential separation and estrangement from nature, a potential upon which modernity was able to build. Thus it appears that if we criticise modern societies for being estranged from nature, we presuppose that humans are somehow special and different from the rest of nature. In that case, our criticism of modern societies loses some of its purchase, since complete human embeddedness within nature now looks impossible anyway.

Suppose, then, that we insist upon the contrary view that we are entirely natural, a part of nature and nothing more. If this is so, our culture-making activities must express our natural powers, and therefore any culture that we create must itself be an expression and outgrowth of nature – even in its ostensibly most artificial manifestations such as motorways or nuclear power stations (see Vogel 1998: 171). However, if we are also assuming that being natural is *prima facie* valuable (such that being cut off from nature counts as disvaluable), then these manifestations must be judged benign after all. Indeed, from such a perspective, Freya Mathews has suggested that, since even bulldozers and missiles are part of the natural order, we ought to respect and sacralise them (Mathews 1994: 164–5). Moreover, if these things, and our acts of creating them, are entirely natural, then this might imply that the existence of these things is inevitable – having been determined by the series of processes and causal interactions of which nature consists.

We thus arrive at the following dilemma: to claim that modern cultures are cut off from nature is, apparently, to accept an anthropocentric assumption that human beings inherently exist at a level of distance from nature; but if we reject this anthropocentric assumption, then we seem to lack the conceptual resources to describe our culture as being estranged from nature in a way that contributes to environmental crisis. Is it possible to develop a non-dichotomous view of humans and world while avoiding this dilemma?

In this chapter, I explore a way out of this dilemma which can be found in the writings of the early German Romantic poet and thinker Friedrich Hölderlin.[1] Hölderlin is best known for his poetry, but he also produced various short theoretical writings and fragments during the 1790s and early 1800s. The human relation with nature is a central preoccupation in these writings. Hölderlin suggests that human beings, and human culture, are entirely part of nature, not separate from nature in any respect. Yet he also wants to hold onto the thought that human beings, and the culture(s) that they produce, do enter into opposition to nature (which, translated into a present-day context, would include our routinely acting in heedless or damaging ways towards natural environments). Hölderlin tries to hold these thoughts together by supposing that nature is divided *within itself*: it has split into its human and cultural part and into a contrasting 'natural' part to which the human and cultural part opposes itself. Thus, Hölderlin understands 'nature' in two different

[1] Classifying Hölderlin as 'Romantic' is somewhat controversial. Dieter Henrich claims that Hölderlin was not a Romantic because he did not subscribe to the Romantic theory of fragmentary, ironic literature, but instead 'advocated perfection in the formal composition of poems' (Henrich 2003: 227). However, I treat Hölderlin as a Romantic in the broader sense that he subscribes to an organic metaphysics, as did most other Early German Romantics.

senses. First, as Nature (capital N), the absolute unity that divides itself into human culture and nature. Second, as nature in this latter (lower-case 'n') sense, which consists of a realm of material objects, events, and processes that stands in contrast to human culture as a realm of conscious, rule-governed activities.

In what follows, I elucidate Hölderlin's view that nature divides within itself and ask how successfully this view escapes our initial dilemma. I also question the usefulness of Hölderlin's view from the perspective of our apparently pressing need to find solutions to environmental crisis. This question arises because the surprising and perhaps troubling implication of Hölderlin's view of self-dividing nature is that our most appropriate response to our culture's entrenched antagonism to nature is to do nothing.

5.1 Hölderlin on Self-Dividing Nature

Hölderlin's views on nature belong within the broader philosophical position that he develops in his theoretical writings and fragments, and which also finds expression in his literary output. Dieter Henrich (a leading scholar of German Idealism, early German Romanticism and their intellectual contexts) describes this position as 'unification-philosophy' (*Vereinigungsphilosophie*). The basic idea informing this philosophy is that there was an original unity which encompassed all that exists, including human beings; but that this unity is now lost, so that one of the overwhelming drives that human beings have is to regain this unity – to reunite with other people and with the natural world.

The basis for this unification-philosophy is set out most clearly in one of Hölderlin's early fragments, 'Judgement and Being' (*Urteil und Sein*) of 1795. Here (see Hölderlin 2003: 191) he claims that judgement (*Urteil*) separates subject and object. He means that conscious experience involves the subject making judgements about objects, from which that subject implicitly distinguishes itself as the one judging. In making this claim Hölderlin is reprising, and modifying, Kant's argument that self-consciousness is the condition of the possibility of experience. It is therefore worth recalling briefly Kant's account, from his *Critique of Pure Reason* of 1781/1787 (Kant 2003). For Kant, to have experience I must make judgements about the materials that are given to me in sensation. That is, I must unite these materials under concepts of objects, of which I judge those materials to be aspects or properties. Thus, a precondition of experience is that I *actively* judge, actively synthesise the givens provided by sensibility. For this to be possible I must have an implicit awareness (that is, I must in principle be able to become conscious) of my self (a) as the *agent* of my judging activity and (b) as *unitary* such that my self can confer unity on the materials about which it judges. Moreover, I must be implicitly aware of my self under a particular description, as David Carr (1999: 36–7) puts it in his reconstruction of Kant's argument. In short, I must be aware of myself (c) as thinking (of the objects towards which my judgements are directed) – that is, as an intelligence – and (d) as spontaneous – for, since judging is active rather than receptive, in judging I must be acting not from natural causality but from reason, following self-legislated norms for how one should and must think in order to have experience. In sum, for Kant the precondition of conscious experience is that I must implicitly conceive of myself as a *subject*, a unitary being that is both intelligent and autonomous. But because I cannot experience or, consequently, know my self, even though I must implicitly *think* of myself as a spontaneous intelligent subject, I cannot know whether my self really is such a being. Equally, I cannot know that it is not – and this creates the space for Kant to argue in his moral philosophy that the fact of my subjection to moral obligations reveals that I am an autonomous and rational subject.

Hölderlin's response to Kant's arguments is that the self who is judging *cannot* (actually or possibly) be a subject. Why not? On Kant's own account, Hölderlin reasons, I am able to think of myself as a subject only insofar as I relate thinkingly to objects from which I implicitly distinguish myself as the one thinking about them. My concept of myself as a subject is strictly correlative to my concept of objects. However, to really *be* a subject, I would have to be so at a point logically prior to the existence of any objects insofar as objects are only constituted by the judging activity in which I as a subject engage. At that point, though, no contrast between my subjectivity and objects can obtain, and since subjectivity is only intelligible through that contrast, it is not an intelligible possibility for there to be a subject in existence at that point.[2]

Given that the spontaneous intelligent activity that is the precondition of my experience of objects cannot be executed by any subject, who then is its agent? It cannot be an object, since objects are correlative to subjects. This leaves only one candidate in Hölderlin's view – 'being' (*Sein*), defined precisely as that from which any subject-object contrast is absent, therefore as absolute unity. '*Being,* – expresses the combination of subject and object. Where subject and object simply are, and not just partially, united ... only there and nowhere else can there be talk of being as such' (Hölderlin 2003: 191). However, since being is not a subject, its judging activity cannot be primarily an intellectual process. Instead, it must be a *metaphysical* or ontological process. It must be the process of actually splitting (*trennen*) into subjects over against objects and thereby constituting both subjects and objects as existents. On these grounds, Hölderlin claims that judgement (*Urteil*) 'in the highest and strictest sense' (191) is actually an ontological activity of separating: originally separating, *Ur-teilen* (an insight that he erroneously takes to be confirmed by etymology). Thus 'judgement ... is the original [*ursprünglich*] separation of ... object and subject, ... that separation which makes object and subject first possible' (191–2).

[2] Holderlin does not state his reasoning as explicitly as this. I am indebted to the reconstruction provided by Henrich (1997). Frederick Beiser also reconstructs the argument (without specific reference to Hölderlin, although he attributes it to those he calls the 'absolute idealists' – Friedrich Schlegel, Novalis, Schelling – amongst whom he numbers Hölderlin) as follows: 'the very concept of subjectivity makes sense only in contrast to objectivity, a contrast that ... works only within experience' (Beiser 2002: 152).

Effectively, Hölderlin has converted Kant's account of self-consciousness into a metaphysical theory in which the universe consists of unitary being that is engaged in a process of self-differentiation. Reality as it is in itself, prior to its constitution of the finite objects that appear to us, is one unitary whole, namely 'being' – that is to say, 'being in the true sense of the word' (Hölderlin 1970: I.558). However, Hölderlin's theory also implies that we can know nothing about being (beyond the fact that it exists as unitary), because we are one of the products of the split within it. Indeed, he states that this 'blessed unity, being, cannot be known' (I.558). We always find ourselves ranged against the objects which are the other products of this split. We can only ever be conscious of objects, not of being.

Hölderlin equates this unitary being with nature. We might think that he is simply following Kant's definition of nature as the totality of objects of experience. Hölderlin would then be translating Kant's definition into the view that nature is the totality of all that is, namely, the whole that divides so as to constitute – and encompass – all existents. In that case, Hölderlin would simply be using 'nature' to mean 'reality as a whole', not specifically that part of reality that is non-artificial (or relatively non-artificial). However, Hölderlin *does* also mean nature in this latter sense. That is to say, he not only distinguishes between Nature as original unity and nature as the sum-total of finite physical objects, but he also regards both Nature and nature as 'natural' in the further sense that they are non-artificial.

First, for Hölderlin, neither finite conscious subjects nor the objects of their consciousness can ultimately differ from being. If they did, then being would not be entirely unitary as (*ex hypothesi*) it is. Being, Hölderlin says, is 'wonderfully all-present' – there are no existents in which it does not manifest itself (Hölderlin 1994: 394). Being must therefore produce the plurality of finite subjects and objects as parts of itself. However, subjects and objects also differ from being, because they are finite and different from one another, whereas being differs from nothing else, since it is everything. Subjects and objects must therefore be *differentiae* that being produces within itself, dividing so as to become internally differentiated. The fact that being gives itself this complex articulation means that it acts as organisms do, for an organism, according to the early German Romantics, is a unity that organises itself into a system of different but functionally interrelated organs. Still, this only suggests that for Hölderlin, it makes sense to equate being with a kind of gigantic, macrocosmic organism. Does this justify the equation of being and *nature*? Potentially it does. Perhaps things count as natural when – or in proportion as – they grow and develop from within themselves rather than (as in the case of artefacts) taking shape according to plans conceived and imposed upon them from the exterior by others (a contrast between the natural and the artefactual that, we will see presently, Hölderlin holds).[3] If so, then, since being develops and differentiates itself from within, it follows that being is natural (and indeed might be equated with nature as the principle of everything that is natural). Thus, for Hölderlin, although being as nature encompasses all that is finite, whether artificial

[3] Confusingly, however, Hölderlin calls the natural 'aorgic' and the artificial 'organic' (see below).

or natural, being itself is not only the whole of reality but also is itself natural in this specific sense of growing from within itself.

Second, Hölderlin equates original being with nature because he relies on an ancient Greek understanding of nature as *physis*, where *physis* is itself understood to be that which 'loves to hide' or 'likes to dissimulate itself' in the words of the pre-Socratic philosopher Heraclitus. Because being is always disappearing, receding behind the separation between objects and subjects, being can be said to 'love to hide', and so can be identified with nature.[4] Perhaps these are not really two distinct points. Plausibly, Hölderlin is reinterpreting 'loving to hide' *as* self-differentiating. In other words, nature loves to conceal itself in that it never remains in its original unitary state but always disperses itself into the separate entities behind which it then lies concealed.

So far, we have seen that, for Hölderlin, reality as a whole is an organic unity which divides itself into the plurality of finite subjects and finite objects. This unity is Nature as the original absolute unity as opposed to nature, the realm of finite objects of which finite subjects have experience. Both subjects and natural objects are aspects of Nature in this expanded sense. However, this might suggest that not only finite natural objects but also finite subjects are natural, since they too are manifestations of Nature. Yet Hölderlin does not think so. Instead, he regards finite subjects as being *opposed* to both nature and Nature. This might seem self-contradictory. How can he consistently claim that we issue from and remain part of Nature *and* that we oppose ourselves to and are separate from Nature (and nature)? Surely if we are part of Nature, this opposition can only ever be apparent and not real?

Hölderlin, however, wants to insist on the reality of the division, and that Nature really divides itself even as it remains whole. That Nature as a unity must include differences within it, differences that we must acknowledge to be no less real than unity, is a difficult metaphysical position to maintain. A contrasting position is that of Schelling in his identity philosophy period, when he maintained that the Absolute (his equivalent of Hölderlin's Being) alone is real and differences are merely an illusory function of the inadequate standpoint of reflection. Hegel sought to articulate and defend the contrary position – that of Hölderlin – by defining true identity as the identity of identity and difference. If identity does not include difference *as* real difference, Hegel maintains, then it is not truly identity at all but is still differentiated from something else (be it only the sphere of illusion).[5] Whether or not we think that Hölderlin's metaphysics is ultimately coherent, it is because he adopts this difficult metaphysical position that he can resolve our initial dilemma, which he does, effectively, by recognising both its sides.

---

[4] Heidegger says that 'The essence of Holderlin's word "nature" resounds … following the concealed truth of the primordial fundamental word *physis*' (Heidegger 2000: 79). See also Dastur (2000: 87).

[5] For discussion of these views, see Nassar (2011).

5.2 Culture and Nature

As we have seen, for Hölderlin, to be conscious – that is, to have experience of a kind that is organised by judgments – is to experience oneself as different from non-conscious objects, as a subject *versus* a world of objects. For Hölderlin, it is only human beings who are conscious and whose experiences are regulated by rule-governed epistemic judgements. Being set over against objects as we are, we are inherently estranged from these finite objects and consequently, too, from the original unity of Nature. Hölderlin's character Hyperion laments: 'an instant of reflection hurls me down. I reflect, and … the world in its eternal oneness is gone; nature closes her arms, and I stand like an alien before her and do not understand her' (Hölderlin 1990: 4).

Human subjects, on this view, are not simply embedded in their natural environments as are natural objects or creatures that are presumed to lack subjectivity. Human beings are inherently both related to and separate from their natural environments – separate in virtue of their inherent consciousness and estrangement; related insofar as their subjectivity goes hand-in-hand with their conscious relatedness to natural objects. Even so, this does not mean that human beings are special *vis-à-vis* nature, or have any supernatural properties and capacities in virtue of which they are conscious and autonomous. Human beings are separate from and related to nature because they are part of Nature, which divides into these two halves.

For Hölderlin, it is our condition of estrangement from Nature that makes us into distinctively cultural, historical beings. The different cultures that human beings have produced in the course of history are, fundamentally, the expressions of different ways in which they have negotiated their separation from Nature. More precisely, for Hölderlin we have two basic drives: to reunite with Nature and to remain separate from it. Different cultures represent different ways of balancing or negotiating between these two drives.

Why do we have a drive to reunite with Nature? Despite the fact that Nature as original unity is unavailable to us as conscious beings, Hölderlin believes that we retain an obscure feeling, a presentiment, that Nature is our basis and origin, and this fills us with a nostalgia to return into it. 'To reunite ourselves with nature, with a unique infinite totality, is the goal of all our strivings', he says in the preface to the penultimate draft of his novel *Hyperion* (Hölderlin 1970: I.558). From the perspective of this drive, we try to reunite with natural objects and thereby to return into Nature as the original subject-object unity. However, we also have a contrary impulse to maintain our separation from Nature. The origins of this drive remain somewhat obscure. Perhaps it arises because what we are as human beings is beings who are separated from Nature and hence, in the name of realising our essence, we seek to remain separate. *Hyperion*, written from 1792 onwards and published between 1797 and 1799, traces the protagonist's efforts to balance these impulses, reflecting Hölderlin's position that the task of narrative literature is to trace the human effort to reconcile life's conflicting tendencies.

Within his many writings on the topic of how precisely poetic literature is to trace this human effort, Hölderlin sought especially to understand ancient Greek tragedy. His thoughts on this topic reveal how he came to understand human culture, as a whole, in its complex relation to Nature. Ancient Greek tragedies, he suggests, depict a hero or heroine striving to regain unity with Nature within the setting of a culture that is highly separated from Nature. Thus Hölderlin does not straightforwardly endorse the view, promoted by influential contemporaries of his such as Friedrich Schiller, that the Greeks were profoundly close to nature (although we will see that Hölderlin does retain a qualified version of that view). Rather, Hölderlin says, any tragic narrative begins with a pronounced opposition between what he calls the 'organic' *(organisch)* and the 'aorgic' (*aorgisch*). This contrast is not between the living and the non-living but between 'the organized, reflected principle of the spirit and of art … [and] the unreflexive, unrepresented, disorganizing manifestation of nature' (as Hölderlin's translator Thomas Pfau puts it in Hölderlin 1988: 168). The 'organic' is culturally organised in light of conscious human activity, while the 'aorgic' is purely natural and has not succumbed to this organising activity.

Hölderlin argues that in ancient tragedies, the hero or heroine strives to overcome this opposition and unite culture with Nature. However, the hero or heroine can only ever bring about this unification in transitory and unsatisfactory ways. For example, Empedocles – the hero of Hölderlin's own tragedy *The Death of Empedocles* – endeavours to live in relative harmony with Nature in comparison to his surrounding culture, but he inevitably becomes compelled to try to reform that culture, which means that he has come to behave artistically – and artificially – *vis-à-vis* that culture, trying to remould what exists in line with ideals of his own that are external to it. Empedocles thereby comes, despite himself, to embody the antagonism of culture to Nature, rather than embodying their unity as he initially sought to do. To escape this problem, Empedocles withdraws completely from the social world, and in the end his desire for culture-Nature unity can only assume the form of the suicidal urge to fling himself into Mount Etna. That is, basically, Empedocles finds that he can only reunite culture with Nature by ceasing to exist as a conscious being at all.

In my exposition so far, it has seemed as if Hölderlin believes in a general, trans-historical opposition between conscious humanity and Nature, an opposition that only Greek culture expresses. However, this is a simplification. While working on *The Death of Empedocles*, Hölderlin formed an account of the difference between ancient Greek and later European culture. In later drafts of the tragedy, he contrasts Empedocles to his 'opponent', a character who personifies what Hölderlin came to identify as the distinctively 'Hesperian' – post-Grecian European – attitude. Hölderlin's 'Hesperian' accepts the suffering, the 'living death', attendant upon our separation from Nature. He writes that 'in his [Empedocles'] opponent … the subjective takes on more the passive shape of suffering, endurance, firmness' (Hölderlin 1988: 61). Rather than striving beyond existing forms of life for unity with nature, post-classical Europeans stoically endure separation from Nature. The same contrast appears in Hölderlin's 1803 'Remarks on Antigone', in which the Hesperians are said to inhabit a culture that 'forces the eternally anti-human course of nature … more decidedly down onto earth' (1988: 113). Our impulse to reunite with nature is more repressed.

Hölderlin, then, came to see Greek tragedy as expressing a culture whose members strove for unity with Nature. How does this fit with his initial view that the Greeks were highly distanced from Nature? Intriguingly, he claims in his letter to his publisher Wilmans of September 1803 that it is because the Greeks had such a strong desire to reunite with Nature, and constantly risked being overpowered by this desire, that they imposed upon themselves and upon their artistic products an excessively severe and austere form, not allowing their desire for unity to find cultural expression. They distanced themselves from Nature *because* they were so driven to reunite with it. In contrast, we 'Hesperians' are much more firmly separated from Nature. The classical effort to achieve distance from Nature has become habitual to us, ingrained, which equally means that our drive to reunite has been rendered comparatively weak. For this very reason, Hölderlin suggests that we can tolerate artistic works that are much more unrestrained than classical works in their level of physical and visceral concreteness, their level of proximity to natural forces and to the orgiastic exuberance of the aorgic, which Nietzsche would later name Dionysian.

This shows how, for Hölderlin, different cultures express different ways of negotiating our dual impulses to reunite with Nature and to remain separate. The Greeks, with a very strong desire to reunite, had to constrict that desire tightly (particularly through tragedy, in which heroic characters are seen giving that desire expression only to be destroyed by it – confirming again the necessity that the desire be constricted). In contrast, we post-classical Europeans are much more firmly separated from Nature, and so we can be relatively relaxed about expressing our desire to reunite with Nature, since this desire is comparatively unthreatening to our cultural life.

We are now in a position to clarify why Hölderlin did not regard both subjects and objects, *qua* parts of Nature, as equally natural. There are two reasons for this. First, we human subjects are estranged from Nature and have some awareness of this estrangement; we necessarily experience ourselves as standing apart from Nature. Second, in negotiating our relation to Nature, we necessarily produce works of culture, works that are artificial or in Hölderlin's idiosyncratic language '*organisch*': their principle of generation is outside them, in their authors. Collectively these works constitute overarching cultures which educate and shape us as human beings. Our distinctive character is to surround ourselves with artificial, culturally created, environments, so in this respect too we cannot straightforwardly be described as natural. Yet the question returns: are not our exercises of intelligent spontaneity, through which we create cultural works and entire cultures, still manifestations of Nature and therefore natural? Not entirely, from Hölderlin's perspective, because these exercises result from our confrontation with our condition of *estrangement* from Nature.

On the other hand, our condition of estrangement from Nature is *itself* natural, resulting from Nature's internal self-differentiation. Human beings, human cultures, and the human activity of producing cultures, thus both are and are not natural: they reflect the self-division or self-splitting that takes place within Nature. They are a part of Nature, but one through which Nature turns against itself and introduces division into itself.

How satisfactorily does this view that human cultures represent an internal division within Nature address the dilemma with which I began? From Hölderlin's perspective it is, at least, coherent to say both that cultures can be and often are estranged and separated from Nature and that these cultures are natural, because they are part of an internally divided Nature. Can some cultures be more cut off from Nature than others? Again, the answer is 'yes'. Indeed, Hölderlin identifies modern Europe specifically as a culture that is particularly entrenched in its position of separation from Nature. It would therefore be unsurprising if modern European – and more broadly 'Western' societies – involved forms of life that are particularly environmentally destructive.

However, it might be said that Hölderlin solves the initial dilemma at the cost of introducing conclusions which are themselves unacceptable. In particular, his position seems to imply that our best response to environmental crisis is to adopt a totally quietist stance: to do nothing. I address this issue in the following section.

5.3 Hölderlin's Environmentalist Quietism

We have seen how while working on his *Empedocles* tragedy Hölderlin formulated a contrast between two possible responses to separation from Nature: the Greek response, to seek reunification; and the 'Hesperian', to accept separation. Hölderlin seems ultimately to favour the Hesperian outlook. He never systematically explains why, but we can reconstruct why he would have come to find it more appropriate given the logic of his thinking concerning Nature.

Since humanity was originally wholly absorbed into Nature, its separation from Nature has not arisen from any activity on humanity's part. Rather, this separation has arisen *from Nature*. If we (human beings) were to interpret ourselves as being responsible for our division from Nature – and to blame ourselves for our antagonistic actions towards nature – then ironically we would actually be presuming that we are capable of acting independently of Nature, assuming that we are the authors of our own antagonistic stance. In reality, though, as Hölderlin says in a letter to his brother of 4 June 1799, 'all the … streams of human activity … have their source in nature' (Hölderlin 1970: II.820). It is not our fault that we are antagonistic towards nature. This is a stance that is imposed on us by Nature, and which is inherent in our character as human beings, that is, as the divided part of Nature.

From Hölderlin's perspective, then, to blame ourselves for being antagonistic towards nature is to make falsely anthropocentric assumptions. Similarly, it would be inappropriate for us to try through voluntary action to overcome our separation from Nature. To suppose that we can overcome this separation is to assume that we can act independently of Nature and can undo the division which Nature has created within itself. This, again, is a falsely anthropocentric assumption. In reality, we have become separated from Nature by its power alone, so it is not within our power to undo separation. Perhaps Hölderlin's greatest statement of this thought occurs in his poem 'As on a holiday …', where the poet attempts to precipitate reconciliation with Nature on his own initiative and is cast down, humiliated, by the gods in punishment for this *hubris* (see Hölderlin 1994: 399). There is also a historical dimension to these thoughts: that the classical Greeks, or their iconic figures such as Empedocles, did commit this *hubris*, and that post-classical Europe represents a progression insofar as it adopts a more modest stance.

In thus endorsing the Hesperian attitude, Hölderlin appears to have ceased to be critical of destructive modern practices towards nature. For, apparently, we are to remain in a state of separation from nature, such that it is always possible for us to act destructively and such that this possibility is rooted in our fundamental condition. In our separation from nature we feel ourselves to be actually opposed to natural objects, a feeling that can always develop into practical self-opposition to nature. What then should we 'do' in the face of this destructiveness and the crisis that it has precipitated? From Hölderlin's perspective, the only appropriate response is to recognise our dependence on Nature and hence to endure our separation from it. If we are unhappy about this separation, then we can only hope that *Nature* may change its mode of being – may at some point re-attain unity with itself. It is our role to wait, patiently and hopefully, until this epoch-changing event may occur. Hölderlin sees this waiting in terms of our need to accept the suffering that falls to *us*, our exile from the original 'peace' of Nature. Today, it would seem rather that our division from Nature spells disruption of nature's own regularities. Nonetheless, Hölderlin's conclusion would remain the same: to try to act differently towards nature would be to make the anthropocentric assumption that we have sovereign authority over our own conduct towards nature, which we do not. Ultimately, contemporary environmental crisis is produced by Nature – through humanity, but not through any sovereign agency on humanity's part. Having no such sovereign agency, we should adopt a quietist stance in the face of this crisis.

However, we need to qualify Hölderlin's claims, for he does think that we can valuably prepare ourselves for a possible change in Nature's mode of being. Poets play a key role here. The poet helps to prepare us for this possible event by evoking the original unity of Nature and by lamenting the passing of Nature's unity. Equally, what for Hölderlin amounts to the same, the poet laments the disappearance of the gods from the world. For Hölderlin, the ancient belief in a pantheon of gods reflected a belief that natural forces are divine powers governing human affairs. Thus insofar as we no longer hold this belief, we no longer see ourselves as part of Nature. The disappearance of the gods from the world indexes our deepened alienation from Nature and its deepened internal division. The poet also celebrates in hymns an imagined future reconciliation between humanity and nature, and anticipates a time when humanity's domination over nature will become submerged by Nature as it re-attains unity. In these various ways, the poet puts us all in an appropriate frame of mind to receive a change in Nature's mode of being should one occur.

With this qualification in mind, is Hölderlin's quietism totally unhelpful with respect to our apparently urgent need to try to act in ways that might avert or lessen environmental crisis? Perhaps not totally. First, we should recall that Hölderlin's quietism is motivated by considerations that count as environmentalist (or, to avoid anachronism, proto-environmentalist): namely, his concern to re-emphasise that human beings cannot act independently of Nature. Second, Hölderlin's position anticipates more recent worries that seeking a 'technological fix' (currently, for instance, in geo-engineering) perpetuates the same mind-set that bears responsibility for environmental crisis. In particular, seeking a technological fix involves looking for ways to use objects to achieve goals defined by us, as human beings – thus presupposing our sovereign agency again. In fact these recent worries about technology stem indirectly from Hölderlin, by way of Heidegger, whose hugely influential arguments about technology derive their structure from Hölderlin.

For Heidegger in 'The Question Concerning Technology', originally published in 1954 (see Heidegger 1993), technology is a way of revealing what is as either resources for human use or as the human agents who are to use these resources. However, for Heidegger, all this is the prevailing way in which Being has elected to reveal *itself*. Technology is not a human creation – to think that it is would be to remain within the technological way of looking at things, rather than asking how technology, as a particular configuration of human-nature relations, is possible at all. It follows that we, as human agents, cannot put an end to the reign of the technological way of revealing, and to try to do so is only to buy into the technological view of our own special, agentic status and thus remain subject to technology after all. Heidegger concludes his meditations on technology by quoting Hölderlin: 'but where danger is, there/the healing power grows also … ' (Heidegger 1993: 316). The healing power is Being itself – currently revealing itself as technology, the danger; but with this comes the possibility that Being might begin to reveal itself in a new, non-technological way. The structure of this analysis clearly comes from Hölderlin.

A third way in which Hölderlin's quietism need not be wholly unhelpful is that our preparation for possible change on Nature's part might include many activities usually seen as being likely to mitigate environmental crisis (such mundane activities as composting, recycling, walking more and driving less, buying local produce, maintaining the fabric of the environments around us, etc.). Ordinarily, these activities are justified by their causal consequences: in scientific terms, as being likely to alleviate or not worsen environmental crisis; and in metaphysical terms, as bringing about or helping to bring about human-nature reconciliation. From Hölderlin's perspective, these activities might be re-considered as ones through which we practise an everyday respect for and courtesy towards nature and thus train ourselves to be as non-antagonistic to nature as we can within the framework of our entrenched estrangement. In this way, we can be expressing and developing our state of preparedness for a possible re-unification with nature – our willingness to relinquish our condition of estrangement should the occasion arise. In that case these activities are justified not just in terms of their causal consequences, but also insofar as they express our preparation for Nature to change. The causal consequences of these activities figure into this expressiveness, since it is because our activities can be expected to have beneficial consequences (in scientific terms) that our doing them expresses our openness to reconciliation with nature. All the same, it is in terms of the expression that these activities are ultimately justified.

The merit of Hölderlin's position is to recognise both the metaphysical reality of human embeddedness in Nature *and* the historical reality and deeply entrenched character of our separation from Nature. Moreover, the quietist consequences of this position need not be as unhelpful as it initially seemed. Rather, we can interpret Hölderlin's quietism in such a way that it justifies everyday, practical, environmentalist activities on the grounds of the attitude they express: recognition and affirmation of our dependence upon Nature.

References

Beiser, F. (2002). *German idealism: The struggle against subjectivism 1781–1801*. Cambridge: Harvard University Press.

Carr, D. (1999). *The paradox of subjectivity*. Oxford: Oxford University Press.

Dastur, F. (2000). Tragedy and speculation. In M. de Beistegui & S. Sparks (Eds.), *Philosophy and tragedy* (pp. 78–87). London: Routledge.

Heidegger, M. (1993). The question concerning technology. In D. Farrell Krell (Ed.), *Basic writings* (pp. 283–318). New York: Harper Collins.

Heidegger, M. (2000). *Elucidations of Hölderlin's poetry* (Keith Hoeller, Trans.). New York: Humanity Books.

Henrich, D. (1997). Hölderlin on judgment and being. In E. Forster (Ed.), *The course of remembrance and other essays* (pp. 71–89). Stanford: Stanford University Press.

Henrich, D. (2003). *Between Kant and Hegel*. Cambridge: Harvard University Press.

Hölderlin, F. (1970). In G. Mieth (Ed.), *Sämtliche Werke und Briefe* (Vol. 2). Munich: Carl Hanser.

Hölderlin, F. (1988). *Essays and letters on theory* (T. Pfau, Trans.). Albany: SUNY Press.

Hölderlin, F. (1990). In E. L. Santner (Ed.), *Hyperion and selected poems*. New York: Continuum.

Hölderlin, F. (1994). *Poems and fragments* (M. Hamburger, Trans.). London: Anvil Press.

Hölderlin, F. (2003). Being judgement possibility. In J. M. Bernstein (Ed.), *Classic and romantic German aesthetics* (pp. 191–192). Cambridge: Cambridge University Press.

Kant, I (2003). *Critique of pure reason*. (N. Kemp Smith, Trans.). London: Macmillan.

Mathews, F. (1994). Relating to nature. *The Trumpeter, 11*, 159–166.

Nassar, D. T. (2011). The absolute in German idealism and romanticism. In A. Stone (Ed.), *The Edinburgh critical history of nineteenth-century philosophy* (pp. 29–46). Edinburgh: Edinburgh University Press.

Vogel, S. (1998). Nature as origin and difference: On environmental philosophy and continental thought. *Philosophy Today, 41*, 169–181.

Chapter 6
Toward History and the Creaturely: Language and the Intertextual Literary Value Space in Jonathan Safran Foer's *Eating Animals*

Tom Bristow

Which fuses me into you now, and pours my meaning into you?
Who was to know what should come home to me?

– Walt Whitman

I'll trade the harvest for the seed

– Joe Pug

Abstract Jonathan Safran Foer's exploration of the language of human-nature engagements critically attends the value space as a site of the exchange of values located in society, transferred to environment and fed back to the individual. Over and above a creative representation of how the environment relates to human values and environmental sensibility, Foer discloses the unexplored areas of our humanity to outline engagements stimulated by a moral outlook which suggests the need for more meaningful relations, those born both from cultural and ecological relations together. He does this through an exploration of what constitutes *the creaturely*. Foer's narratological output to date, consistently figures familial relations as a literary value space wherein the geopolitical context colours idiosyncratic, historically contingent and localized inflections of values. Earlier novels suggest the many ways in which relations are understood in terms beyond value: intimacy, the familial and proximity inhere in a value space that operates beyond the parameters of utility and pleasure. In *Eating Animals* (Foer, J. S. (2009). *Eating animals*. London: Hamish Hamilton) Foer advances these ideas by outlining the relations between language, animality and family in order to deconstruct species barrier dualisms and to move beyond the language of intrinsic value. His

T. Bristow (✉)
School of Arts (E11), LG38, University of New England, Armidale, NSW 2351, Australia
e-mail: tom.bristow@une.edu.au

E. Brady and P. Phemister (eds.), *Human-Environment Relations: Transformative Values in Theory and Practice*, DOI 10.1007/978-94-007-2825-7_6,

text, therefore, highlights the capacities of the imagination to dwell on meaningful ideas of nurturing and of sustainable relations.

Keywords Cultural ecology • Values • Animality • Family • Language

6.1 A Self-Defining Novel

Eating Animals is a curious novel. In fact, very little of the work is novelesque; that is to say, there is no character development and the minimalist plot is hidden. However, it is an engaging narrative in literary prose that exposes the processes of one man's consciousness digesting and translating the issues surrounding factory farming and vegetarianism. I use these terms conspicuously. While reality is not created within the fiction – it is merely represented factually via academic, governmental and industrial reports – it is woven into Foer's personal life, thus connecting *Eating Animals* to his other literary projects that attempt to marry the geopolitical to the familial. Any impasse here must be overcome. *Eating Animals* takes this to the next level through an explicit discursive philosophical rumination on how the narrator-protagonist can best feed his newborn son. In Foer the progenitive is caught in an ethical net of judgement and legacy: evolutionary and ecological dimensions are coloured with cultural relations that range from a spectrum of intimacy (father-son, boy-grandmother) to a collage of disconnections (human-animal, food-environment). The consequences of our actions, metaphorized as the ability to reproduce, operate on many levels (from reproducing statistics to reflecting on the resources and materials of future food production levels in light of current technological trends) and are framed via Foer's grandmother's aphoristic phrase that opens and closes the novel: 'If nothing matters, there's nothing to save' (Foer 2009: 17, 267). This incredibly personal and emotive epoché to the novel's evaluative and at times judgemental dynamics offers reason to cherish things in virtue of their being worldly.[1] The capacity to care over and above mere appreciation of value is central to the novel's allegorical and highly individualistic nature, which is of artistic merit. This said, the most literary moments are those occluded by complex and dense intertextual referencing that require some academic oxygen to bring them to full ethical light. Due to limitations of space, this chapter will unpack these references rather than substantiate ideas within the novel's literary events; intertextual context (historical and ethical) is, therefore, offered as a substitute for literary exegesis.

As the title indicates, the work examines the fact that humans eat animals. Moreover, these animals are themselves fed rendered animals. Thus (as animals ourselves) we eat animals that eat animals. In Foer's radical punctuation it might read something like 'EA [EA [EA]]'. Furthermore, as title and paratextual device to frame our thinking, the naming of ourselves within the object of inquiry signifies

[1] See Holland's use of Attfield's definition of intrinsic value in this volume (2011).

the nominative act beyond us, with the other incorporated into the same realm of concern. Here, the verbal noun 'the action or habit of taking food,' is conflated with the participial adjective (gerund), 'that eats' (OED). Not only are we left oscillating between action and description, between process and reflective analysis (or the transitive and constative vehicles of thought),[2] but we are given the first indication that this troubled semantic realm is not only a record of how we forge bonds (through the process of eating together) but is also indicative of the fact that a lack of intelligence or clarity about our actions is eating away at our creaturely selves. Foer's literary value space is designed to regain this sense of ourselves. I examine how the novel achieves this in part by moving through semantic chaos or an unfathomable bricolage of data on factory farming. With his son very much at the centre of his thoughts, this corroding of humanism necessarily, in Foer, relates to the notion of offspring; it is a literary progenitive contour. In Foer's novels, subjectivity, agency and the constitutive nature of discourse are always under scrutiny; in *Eating Animals*: 'Food, family, and memory are primordially linked' (Foer 2009: 194). This is complex and is best instanced in Foer's use of Walter Benjamin. To prepare for discussion of this, I will first inspect more closely this confusion and the corruption of our language and of a contemporary sense of human animals.

6.2 Towards a Definition of Animals

The third named section to *Eating Animals*, 'Words Meaning', constitutes an alphabetised glossary of terms, which includes entries for 'Animal,' 'Intelligence', 'Human' and 'Suffering' among others. The section's title page is footnoted with the following, 'Animal agriculture makes a 40% greater contribution to global warming than all transportation in the world combined; it is the number one cause of climate change' (Foer 2009: 43).

The number has been rounded up from 38%.[3] The gloss for 'Environmentalism' enfolds this and moves cold statistical reporting toward a creative, affective arena:

> Animal agriculture is responsible for 37 percent of anthropogenic methane, which offers twenty-three times the global warming potential (GWP) of CO_2, as well as 65 percent of anthropogenic nitrous oxide, which provides a staggering 296 times of the GWP of CO_2. The most current data even quantifies the role of diet: omnivores contribute seven times the volume of greenhouse gases that [sic.] vegans do. (Foer 2009: 58)

Does the connection between agriculture and climate change adumbrate the novel's ardent sense of character development, or virtue, or the eagerly poised moral question of elective vegetarianism or selective omnivory? Or is the neutral tone silently coloured with finger-pointing moral high ground that undermines its

[2] My thoughts are largely influenced by William James (James 1890: 237–248).

[3] The IPCC has reported that transport constitutes 13.1% of greenhouse gas emissions while the UN has reported that the livestock sector is responsible for 18% of greenhouse gas emissions (Foer 2009: 280–281).

objective, scientific rationale? These questions are implicitly threaded throughout the novel, sometimes with the moral issue amplified, sometimes with the scientific positivism qualified and upheld by grassroots activism or documentary reportage. All are devices provoking the reader to take their own position on Foer's narratological technique and to consider the effects of differing rhetorical strategies within grand narratives (Bristow 2010a). What is of particular interest here is that veganism is (morally) empowering. However, Foer is obviously asking us to think of *the cost* of eating animals. This relates strongly to the need to disclose all that entails in eating animals – not only their suffering, the net contribution to climate change, species loss and biodiversity depletion, but also the cost to ourselves both morally and through the results of forgetting where we are placed within these relations. To do this Foer regains a sense of meaning as that which is constructed by humans and yet is not necessarily only historically contingent.

To set off this train of associative thought, Foer turns to the research of anthropologist Tim Ingold in the gloss for 'Animal', the first entry in this 'Words Meaning' section. An inquiry concerning the definition of this complex noun discloses the 'unexplored aspects of *the understanding* of our own humanity' (Ingold 1988; cited in Foer 2009, p. 46; my emphasis). Summarising his own research methodology, Foer (2009) remarks that he regularly found himself in a state of confusion:

> Sometimes my disorientation was the result of the slipperiness of terms like *suffering*, *joy*, and *cruelty*. Sometimes it seemed to be a deliberate effect. Language is never fully trustworthy, but when it comes to eating animals, words are as often used to misdirect and camouflage as they are to communicate. Some words, like *veal*, help us forget what we are actually talking about. Some, like *free-range*, can mislead those whose consciences seek clarification. Some, like *happy*, mean the opposite of what they would seem. And some, like *natural*, mean next to nothing. (Foer 2009: 45)

So it is with the current state of terminology in food labelling endorsed by governments under pressure from the agricultural industry: 'free-range' might only mean up to 33,000 birds in a shed with access to a crack of light or seven birds per square meter (Foer 2009: 61), whereas being free to roam as in a 'natural environment' might mean the genetically engineered chicken has one square foot in which to carry out its entire foreshortened life. However, it is the sense of what is 'natural' that really puzzles Foer, particularly as it suggests normativity and conformity to communal and familial modes of social being. This has been examined by David Harvey who repents the 'tendency in discursive debates to homogenize the category "nature"[…] when it should be regarded as intensely internally variegated – an unparalleled field of difference' (1996: 183; cited Shukin 2009: 15). While wary of uniformity, ecologically-friendly post-structuralist concepts are cast aside by Foer as he moves to an ever increasing synthesis of quantitative data.

As the empirical, documentary evidence builds up against the American agricultural industry's ethically unsound and ecologically flawed practice in the Food and Drugs Administration and the United States Department of Agriculture, compounded by evidence from the American Dietetic Association, the American Livestock Breeds Conservancy, the American Meat Science Association, the

American Medical Association and the American Public Health Association, it is clear that this institutionalised endemic causes pain and suffering. This is a wellbeing feedback loop in Foer's text, for both animals and humans. It is viewed from the position of a grounded individual rather than an omniscient narrator; evidence and response are always offered local context, ranging from slaughterhouse confessions and reportage to reflective discussions with farmers at the dinner table. Furthermore, the semantic mess at the heart of the endemic is analysed as a combination of social conditioning and the global free-market capitalist power geometry whose survival instinct is to interpolate false dichotomies. One of these is the species barrier: 'farmers – *corporations* is the right word – have the power to define cruelty', we are told, vehemently (Foer 2009: 51). This critique has its own view on things delivered in the name of progress. In wishing to deconstruct the dichotomy, Foer heads towards interdependence and feedback-oriented co-creation.

6.3 Ingoldian Ruminations

Foer draws from Ingold's *What is an Animal* (1988) to outline a consensus drawn from a diverse group of scholars. Reflecting on the meeting points of the academic discourses of anthropology, literature, philosophy and theology, Ingold (2008b), during a similar conjunction of ideas some years later, declared that it was necessary to clarify three concepts. First, there are two ways of discussing 'holism': (i) relying on the difference within part-whole relations; (ii) processual conceptions foregrounding the moment and the encapsulation of the whole within a field in time. Second, 'embodiment' is becoming a derivative term, a panacea without definition; a new direction to be considered is that the body needs to be 'enworlded'. The example of kite-flying would offer models of agency and patience (kite and human with string as vector of interaction). However there would be no flying without the wind. Both kite and human are immersed in the current of the medium and as such they are works-in-progress (Ingold 2008b). The first two concepts thus provide an imagined model of human agency and technological incorporation alongside the intermediation between human and world. This is offered as both entertainment and for contemplation when we are mindful of context and the affording structures for our actions and thoughts (see Holland 2011). The third concept follows more philosophically thus:

> The nature/culture divide was resident in the dichotomy between the wild and the artificial, resonant in the terms/metaphors and models used in our presentations: two issues lead from this: (i) if all is cultural or artificial, where is the room for human imagining?; (ii) a new materialism would focus not on material culture but on *the actual materials of our experience* and accordingly, the conditions of possibility for the affordance of such materials (Ingold 2008b; my emphasis).

This materialist view chimes well with Foer's compressed sense of 'Eating Animals' EA [EA [EA]]. Following Ingold's sense of environment and agency, and while inspired by Foer's use of this as energy within his narrative project, I would also say that there is room for entrance into the imaginative and creative interchange between things. This space is not necessarily betweeness as stillness (see discussion of Rilke, below), but it is a place for growth.

Ingold's 'The Environment as Fluid Space' (2008a) articulates the trial or 'movement of growth, which refutes the logic of inversion [that which protects the inward/outward model]'. Foer is keen to see how the past is reconfigured in the present as means to move forward and out of the contingent semantic confusion; space and place for Ingold, however, are no longer temporal co-ordinates but events in a vision of 'the material world precipitating out of materials meeting, of the fluid nexus and flows of life'. This is what he has called '*the parliament of lines*' (2007: 5). Ingold's emphasis (2008b) was exacting: a post-phenomenological stance reads the web and the fabric of life in a new way as 'gatherings' and 'entanglements' – meshwork over network. 'Meshwork' is neither messier nor less sophisticated than 'network': 'network' simply offers a degree of intelligence (computation and structure) that is understated by Ingold's 'meshwork'. It is worth evaluating Foer's text in terms of consonance with this rich, ecological concept.

6.4 Shame, the *Unknown Family* and Forgetting

Citing E.C. Spada, Foer discloses his project as 'the continuous critique of our working definitions in order to provide more adequate answers to our questions and to that embarrassing problem that animals present to us' (Spada 1997; cited in Foer 2009: 47). Foer interprets this as the problem 'that we don't simply project human experience onto animals; we are (and are not) animals' (2009: 47). Much as we are part of and apart from the environment, this problem – traditionally thought through the philosophy of mind-body dualism, intrinsic versus instrumental value argumentation or debated in the problems of ecological terms clarifying metaphysics, especially part-whole relations – acutely clarifies that Foer's sense of anthropomorphism is not only learned and partakes in the Ingoldian philosophical spirit, but it is also self-reflective. To help the reader reflect more on self-reflection, or on reflection as a form of environmental and moral feedback, Foer takes us on a journey into the heart and mind of Kafka.

[Animals and shame]

> Shame is what we feel when we most entirely – yet not entirely – forget social expectations and our obligations to others in favor of immediate gratification. Fish, for Kafka, must have been the very flesh of forgetting: their lives are forgotten in a radical manner that is much less common in our thinking about farmed land animals. (Foer 2009: 37)

As I have indicated, Foer is attempting to posit vegetarianism as empowerment over and above relinquishment. The reminder that our action is at odds with social expectations, although forgotten momentarily, is the very pulse of shame. So it is that 'shame is the work of memory against forgetting' (Foer 2009: 37). In reading

Foer, this social aspect to our feelings about our actions is not powered by guilt in light of witnesses. We learn, but do so through the coordinates of comfort and discomfort with fellow humans, which highlight the need for courage:

> Try changing napkins at Thanksgiving...– even do it bombastically, with a lecture on the immorality of such and such a napkin maker – and you'll have a hard time getting anyone worked up. Raise the question of a vegetarian Thanksgiving, though, and you'll have no problem eliciting strong opinions – at least strong opinions. (Foer 2009: 264)

Ingold has helped Foer to identify the unexplored area of our humanity that is the narratological locus in *Eating Animals*. Foer wants to move around this uncomfortable arena to show how it needs representation to come to full ethical, humanist light. 'The question of eating animals hits chords that resonate deeply with our sense of self – our memories, desires, and values [...it] is ultimately driven by our intuitions' (Foer 2009: 264). As a project, accounting for our intuitions – and by implication our latent values – as something that could inform our working definitions (vis-à-vis Spada), ideas are mobilized to actively bridge the gap between individual and society, rather than to critique a lack of a rational, cognitive point of reflection. This summarizes Foer's advanced literary imagination in *Eating Animals*. It is for this very reason that he turns to Benjamin's commentary on Kafka.

Benjamin (1992) argues that shame is Kafka's strongest gesture and that it has a dual aspect:

> Shame is an intimate human reaction, but at the same time it has social pretensions. Shame is not only shame in the presence of others, but can also be shame one feels for them. Kafka's shame, then, is no more personal than the life and thought which govern it and which he has described as thus 'He does not live for the sake of his own life, he does not think for the sake of his own thought. He feels as though he were living and thinking under the constraint of a family... Because of this unknown family... he cannot be released.' We do not know the make-up of this unknown family, which is composed of human beings and animals. But this much is clear: it is this family that forces Kafka to move cosmic ages in his writings. (1992: 125–126)

Kafka bids for this family and in so doing, he moves 'the mass of historical happenings', in effect placing humans in a new realm and clear space from where one can truly view the 'swamp world' of reality (Benjamin 1992: 126). Foer's redefinition of words without cultural slippage is an attempt to escape the Sisyphean circle of swamp-clarity-swamp, i.e., each instance of clarity moving through political and contingent events that cloud meaning over time. Redefinition and clarity of terms for the contemporary ecological context of our actions, and for a glossary of the animality- humanity language economy, follow Ingold's thread to examine how we understand ourselves as human animals. Herein lies a better understanding of how our actions relate strongly to how we are motivated; or better, are 'driven by our intuitions about what it means to reach an ideal we have named, perhaps incorrectly, "being human"' (Foer 2009: 264).

[Animals and forgetting]

The reader may be perplexed at the reasoning behind Foer's insertion of Benjamin's reading of Kafka during a materialist inquiry into human action. The reader may also be perplexed at the net effect of the shift in rhetorical tenor wherein fish are read as 'the very flesh of forgetting: their lives are forgotten in a radical manner' (Foer 2009: 37). Foer is reminding us of our capacity for moral sensibility. He is informing us that animals are part of a community in which we might feel shame. He turns to Benjamin's understanding of shame to do this as it offers a historical perspective from where we can view the bricolage of broken relationships (and the semantic mess) as a resultant effect of our actions and as a consequence of forgetting.

> Beyond this literal forgetting of animals by eating them, animal bodies were, for Kafka, burdened with the forgetting of all those parts of ourselves we want to forget. If we wish to disavow a part of our nature, we call it our "animal nature." We then repress or conceal that nature, and yet, as Kafka knew better than most, we sometimes wake up and find ourselves, still, only animals. (Foer 2009: 37)

Foer finds it strange and confusing that humans, while capable of recognizing elements of ourselves in animals (he provides the examples of spines and nociceptors in fish), simultaneously deny the value and importance of these similarities. This is a form of forgetting which 'equally den[ies the] important parts of our humanity. What we forget about animals we begin to forget about ourselves' (2009: 37). This forgetfulness constructs a false species barrier while denying our humanism, too. All this is positioned within the novel to 'frame a small story' about Kafka's gaze resting upon the bodies of fish in an aquarium (2009: 36).

Foer references Kafka's friend, Max Brod, recalling that Kafka spoke to the fish: '"Now at last I can look at you in peace, I don't eat you anymore." It was the time that he [Kafka] turned strict vegetarian' (ibid.). This story, morally injected with Foer's understanding of Benjamin's Kafka, has enabled the novelist to speak of the intimate and social aspects of shame. Rather than reading intrinsic value or the value of the fish as exclusively intrinsic, it casts humans as both '*interactive* agents rather than observers… and *creative* in their dealings with the natural world' (Holland 2011; my emphases). Thus, Foer's intertextuality promotes the value of meaningful relations derived from ecological and cultural attunement, motivation and mindfulness. This interactive, creative paradigm resonates with Foer's interest in connecting the familial to the geopolitical. It brings animals within our understanding of material culture as part of those invisible others that are actually '"unknown family,"' to use a phrase from "Er" ("He") (Kafka 1954: 295).[4] They are thus '"receptacles of forgetting"' (Foer 2009: 36). This last phrase is Benjamin's; Foer, however, is not prepared to go the extra ground to illuminate his dark materials. It is useful, as I shall show.

Readers of Benjamin's text will note that Foer has reconfigured agency and object: Benjamin writes of animals as 'receptacles of the forgotten', not 'of forgetting' (1992: 128). Benjamin's lost, unknown family is contained within the animal signifier; Foer's emphasis on the present tense and the process of human thought emphasizes the Benjaminian historical dimension rather than the sense of inclusion (although the two are related). While Benjamin's sense of 'forgotten' resonates with the pile of debris that the Angel of History would like to heal – but cannot for it is blown forth by the wind of progress (1992) – Foer's verbal noun discloses the action

[4] Kafka speaks of a law that negates living and thinking only for oneself.

of the verb to forget *and* the 'state of being unconscious, oblivion' (OED): action and result within the same moment. It is a subtle difference in literary articulations of time, but it does help to think whether there is room for redemption from the past or present in either writer's ethics (I shall briefly address this, shortly). Benjamin is discussing Kafka's surrealism, wherein redemption or hope is not available for those present now, only for others later. Foer, by contrast, has to get us to an understanding of how we can change things for a better future or there would be no point in considering vegetarianism.

To pass on may require one to forget, but that might be to pass over. The grandmother wishes to pass on her values; Foer as father wishes to pass on his findings with respect to food production couched within his desire to pass on a healthy and moral legacy to his son; the latter is symbolically charged with the sense of the depleted biodiversity that this generation will leave behind for future generations. Both of these develop a sense of remembering which comes from the reuse of language as explicitly indicated in Foer's intratextual frame of inheritance and duty – his grandmother's words existing as an echoic material event situated at the novel's beginning and end. Much like hermeneutics is a form of transformation, restatements are acts of cultural production themselves. They are not a parasite on the process of interpretation for they create something new as things are passed on – if only, as in the case of naming the bitch (canine) in *Everything is Illuminated*, 'Sammy Davis, Junior, Junior', to echoically reify the progenitive matrix by going beyond the individual scale.[5] This sense of the need to be passed on and renewed – or to resound via simultaneously conceiving and enacting voice and breath, taking in and giving out: to be *inspired* (Bristow 2010b: 23) – is one implicit element within Benjamin's multi-dimensional ethics of development and progress. In Benjamin own words, 'The language of nature is to be compared to a secret password, which passes on each post the next in its own language, contents of the password' (1996: 74).

This is the penultimate sentence of Benjamin's 'On Language as Such and on the Language of Man' (1916). Paul de Man's reading of Benjamin in the posthumous *Resistance* (1986), while suspicious of intralingual equivalence of signifier and signified, emphasizes the need for the political stance indicated as courage in Foer. As an historic lens, it instances the ripening of the material world through man's actions: '[T]o understand natural changes from the perspective of history, rather than history from the perspective of natural changes' (de Man 1986: 83).

Benjamin's ethics is underwritten by an eschatological mindset that posits history as the framework for our responses to and from within the world. Foer's emphasis on economics and choice, therefore, resonates with de Man's post-Romantic clarification of Benjamin's project. The historical lens attuned to material events via de Man, might open an interface with existential ethics in terms of existence preceding essence, but I argue that Benjamin and Foer think within and outside the

[5] This particular example also reifies the self-enclosed human ego, problematizes fixed gender and incorporates the other (the non-animal human) as one's cultural offspring. Paul Carter (2010) argues that echoic framing – the repetition of some sound shapes and names – contingently incorporates the encounter between peoples and place within the moment of naming

conditions of existence of the individual person; while in Foer, the individual's emotions, actions, responsibilities and thoughts are central to an ethical perspective that he ties to a question of intersubjectivity and the idea of a shared destiny.

To be more precise, Benjamin is speaking of language as the expression of those aspects of mental life that are communicable. We might wish to recall Foer's shame-laden yet potentially progressive table-manners, and think of the potentiality for raised consciousness within the debate on vegetarian Thanksgiving as one that can re-fertilize the historically contingent issues at hand: namely, our use of materials, the materiality of animals, and the cultural make-up of humans. Benjamin is drawing from language in the wide sense of a method of expression, not just as a lexical field. For Benjamin, the human's mental being communicates in language and it follows that it is in the naming process that the essence of his being lies. Rather than claim hierarchical advantage over world-poor animals (cf. Heidegger, below), Benjamin's argument posits: (i) that world communicates to humans *first* by stepping before them, by which we are able to name and go beyond the self; (ii) that a use *in language* rather than by language is how the mental being communicates to God. For Benjamin, naming is this manner, which completes God's creation as a creative power set free in the human (1996: 68). The nominative act can both establish disconnection and distance while it can also broker a relation or common meeting point in understanding. Foer's innovative paratextual devices, outlined earlier, suggest that he has understood some of this. At the very least he has thought about the slippage in language and the (intentional) misnaming of food as an example of the world not communicating to us. In Benjamin's terms this could be the human either not listening to world or not participating in the world's song, not passing on the dynamic properties of the password (keeping the world fertile). Cultural ecologists and literary scholars note that we have entered a critique of the subject that is clarified by attention toward cultural lamentation for meaning and value.

6.5 Time and Mourning

Hanssen (1998: 2) reads Benjamin's efforts to differentiate his writing from Heidegger's as one crystallized in the concept of the *kreatur*, the de-limitation of the human subject within a thesis of natural history that incorporates transience and decay: a non-human history coloured by a critique of the philosophy of the subject. Benjamin's humanistic subject is considered in terms of the present being salvaged by the incomplete potential of the past. This, although aimed against Fascism and anchored in terms of anti-foundational human freedom, critiques morality located in the realm of action ("Fate and Character", 1919) and the politics of destruction ("The Critique of Violence", 1927) as the contact with nature's forces that are violent and revolutionary. Salvation of the present from the incomplete or the as yet unrealized posits non-foundational origin, which purports a new ontology of vitalism where fostering potentiality and openness to change are key. Perhaps the dissonance with Foer on this point is that this is not unlike hypotheses that dispute hereditary mechanisms. Foer's humanistic narrative wishes to foreground learning and the passing on of values – those that are intergenerational and those that are recontextualized; those that are latent or submerged; those that are mistranslated or misnamed. His poetics, conversely, support a sense of rupturing vitalism, newness in nature's autopoeisis. We need something else to make more sense of this duality in Foer, especially as it came from a desire to read animals expansively.

Benjamin's view on the past is his basis for understanding the processes of history, a view coupled to concerns with transience and decay (from which may come growth). Heidegger's historical view, on the other hand, postulates a submerged humanism and suggests that nature should not be contrasted with history (human action) and that greater insight into the possibilities of relation will allow entrance into the boundless where humans and animals can be at peace. This appears to be of more use to us.

On Heidegger's reading of Rainer Maria Rilke's poetry, animals present back to humans the notion of a submerged humanism in the recesses of the heart, the potential of which is to be gained via loss of self. Heidegger argues that the human needs the animal's face (*Antlits*) to connect to the open (the boundless world of unrealized potential that delivers freedom to human consciousness). I read this site as a value space of meaningful relations in which one can act shamelessly in Kafka's sense. In 'What Are Poets For?' (1946) the need for the non-rational is motioned less than the idea of being brought into relation with God via human distinction to creatures. It is thus – in an intuitive realm – that 'nature' in Heidegger's Rilke is not divisional, as it 'is not contrasted to history'. Rather, it is *life*, not in the biological sense, but as *physis*: 'Being in the sense of all beings as a whole' (Heidegger 1971: 101). This might herald true consonance with Ingold's meshwork. Certainly, it is informed by the same ethical premise and philosophical purview. Thus, minding the animals reflects back upon a German-Jewish tradition of the creaturely to emphasize possibilities of reconceptualizing human and non-human relations. For example, in the fourth of the Duino elegies contemplating playfulness and harmony, Rilke asserts that: 'our blood does not forewarn us/like migratory birds' (1989: 4, lines 2–3). The fact that the animals are world-poor suggests that the human, conversely, as world forming, is not absorbed into the environment like the animals (albeit in this line, attuned to invisible energies and paths of sustenance) and thus can view it as a site of possibility from a distance, i.e. from outside. This is a mode of being that is discrete from the phenomenological perspective or experience of oneness.[6] The distinction between *weltbildend* and *weltos* for Heidegger (1995: 273) is that an enormous sadness burdens the latter animal realm. In Rilke's fourth elegy, man is a latecomer to himself, while in the eighth elegy, the human is *backward looking*, his eyes like 'traps' surrounding the

[6] This sense of being apart or not absorbed might inform what Derrida means by humans as '*questioning* entities' (Baker 2006, p. 71). Conversely, 'Derrida's resistance to the philosophical doxa that language constitutes an absolute boundary between animal and human involves identifying animals with the immanent otherness of logos, something he achieves by suggesting the tropological sites of language, specifically metaphor, are animal' (Shukin 2009, p. 33). For direct relevance to Heidegger see Derrida (1991) and Derrida (2002); for full context consult Shukin (2009, pp. 29–42). A more informative application of Heidegger's distinction between the poverty of animals and human openness is offered in Wendy Wheeler's clarification of being spellbound (2010: 46).

world as it 'emerge[s into] freedom' (Rilke 1989: 4, line 4). While not conceptually accurate, this resonates with Benjamin's 'Angel of History' mentioned earlier: the metaphor of fleeing from the womb is the pain or 'enormous sadness' (Rilke 1989: 4, line 42), which, due to *memory of* the first and now distant home, provides the condition that 'we live here, forever taking leave' (Rilke 19891: 4, line 73; my emphasis). I have indicated that this mode of consciousness is separate from the world, outside a position of animality: it is a position of loss. Respite from this mourning and sense of loss comes where there is no 'World', no 'pure/unseparated elements which one breathes/without desire and endlessly knows' (Rilke 1989: 4, lines 16–19). This is disinterestedness as 'unuberwachte' (line 18), free without burden of supervision and healed in 'timeless/stillness' (lines 17–18) that relinquishes the goal-oriented and utilitarian vision of the future but enters into the 'boundless,/ unfathomable' (lines 36–37). This is a cousin to Ingoldian meshwork.

I return now to Benjamin's password that is shared between sentinels, a resource or energy that inheres in the text that is to be 'translated'. For something to be shared in this way there must be a web of bodies, an interaction that leads to fertilization. This is a process and a means to interface with others. It is not only a muteness in which things shine forth – a reverse echo of the nameless (nature) into the named (human realm) by promoting the named (human) into the nameless (nature) – but it is also German-Jewish sense of co-creation while dispossessed of world: living with things without territorial ownership and control: continuity and consanguinity over protectorship, stewardship and progenitorship. I locate this as a correlate to Ingold's concept of meshwork, which substantiates his work on animals that informs Foer's cultural critique. In Foer, historical continuity and the potential for an imagined difference in the present do not merely portend that we need to constrain our decisions and actions as means to making an appropriate future for others. They also help exemplify the fact that humans need room for contemplation on their actions, which will give rise to expansive inspiration and aspiration as creative agents (Holland 2011). Our grasp of the meaning of how we receive language and regenerate language, is like the material world it represents, 'related to our understanding of how culture is produced, disseminated, and consumed' (Shaw 2008: 21). It is important to follow this through as the emphasis on human action (as in the work of de Man) suggests particular care with the materials of our existence (Ingold 2008b): how we eat; that we are *eating* animals. Foer's verbal participial inflection to Benjamin's adjectival sense brings the fiction back to the realm of Benjamin's language of nature, rather than to his thoughts on Kafka's fiction. Here, the action, which is a form of dispossession, can bring the redemption or positive change Foer requires.

6.6 Concluding Remarks: The Language of Nature

The framing maxim to *Eating Animals* amplifies the (minimalist) chronological progression in the otherwise plot-less novel. Character is developed through raised awareness of virtue, which comes from incorporating the consciousness of the other (in this instance, Foer's grandmother). The reported grandmother speaks the exact same line twice: 'If nothing matters, there's nothing to save' (Foer 2009: 17,267). In the opening of the novel, this line follows the young Jonathan asking why she refused to eat a piece of meat offered to her by a Russian farmer who saw her starving as she was running from the Nazis: '"You didn't eat it? [...] But not even to save your life"'. The grandmother's response (above) puts life and ethics before individual primary needs. Continuity and contact with previous modes of being and with preceding historical conditions are crucial for Foer's sense of the creaturely and his desire to motivate courage and aspiration in his reader. We visit the line for the second time while the mature Jonathan reflects on his research for the novel:

> Whether I sit at the global table, with my family or with my conscience, the factory farm, for me, doesn't merely appear unreasonable. To accept the factory farm feels inhuman. To accept the factory farm – to feed the food it produces to my family, to support it with my money – would make me less myself, less my grandmother's grandson, less my son's father. (Foer 2009: 267)

This conflation of the idiosyncratic familial with the global and creaturely familial is what Foer claims as the meaning behind the duty of care outlined by his grandmother's maxim. The novel's repetition of the understated theological axiom of charity suggests circularity or stasis. However, the increased sophistication in understanding one's ancestral ethical stance is something that not only relates to our understanding of animals and the crumbling humanism of the twenty-first century, but is also a mitigating force against this crisis—one that understands the (biological and cultural) need to pass on and regenerate in new formations.

I have claimed that Foer's aesthetic bridges the gap between the personal and the geopolitical through an exploration of the language of relations. In *Everything is Illuminated* (Foer 2002a), a young American Jew, called Foer, travels to the Ukraine to seek out the woman who saved his grandfather's life during Nazi liquidation of his family's shtetl. What transpires is an exciting account of how a contingent language community is developed between Foer, his taxi driver, Alex (a Ukranian native who has learned English from a thesaurus without ever hearing it spoken), Alex's blind grandfather and a deranged 'Seeing Eye bitch' (Foer 2002b: 5) called Sammy Davis, Junior, Junior. A variety of performative modes and capacities to see, to speak and to reiterate English are embodied within these characters as *Umwelt* engagement capacities. Foer's narratological connective mixing desk is deployed to route and change the level, timbre and dynamics of human and animal expression. I argue that this is consistent with other of his fictional works that seek to outline the connections between experience and reflection within the language-use of a 'family'. *Extremely Loud and Incredibly Close* (2005) locates a vegan, pacifist, 9 year old boy, Oskar Schell, 2 years after losing his father on 9/11. Schell is linked to the earlier novel's protagonist, Foer, through the paternal grandparents who lived in Dresden. Both texts primarily investigate human-human exchanges across moments in history. *Eating Animals*, looking further down the family tree to Foer's son, investigates these exchanges within the framing context of human-nature relations while remaining precise about the historical context: our contemporary ecological crises coupled to moral bankruptcy. The end of things, however, is less Foer's concern than possible redemption for the future, or, better, his use of creative association clarifies the site in which we can operate with more skill and mindfulness. Foer's intertextual literary value space highlights meaningful relations that are an affordance for more sustainable cultural practice. As a literary event itself, dynamic intertextual interchanges within *Eating Animals* betoken the conditions of imaginative possibility for heightened ecological consciousness. Benjamin has been of great use to Foer in terms of shame, forgetting, and degrees of empathy relating to degrees of care. The deeper bridge between Benjamin and Foer comes when their joint understanding of language, cultural production and cultural consumption is read as being creatively symbiotic.

References

Baker, S. (2006). "You Kill Things to Look at Them": Animal death in contemporary art. In Animal Studies Group (Ed.), *Killing animals* (pp. 69–98). Urbana: University of Illinois Press.
Benjamin, W. (1992). *Illuminations* (H. Zohn, Trans.; H. Arendt, Ed.). London: Fontana Press.
Benjamin, W. (1996). In M. Bullock & M. W. Jennings (Eds.), *Selected writings* (Vol. 1). Cambridge, MA: Harvard University Press.
Bristow, T. (2010a). *Eating animals: Food farms and ethics in contemporary literature.* Paper presented at Universitas 21, University of Nottingham, Ningbo, China.
Bristow, T. (2010b). Phenomenology, history, biosemiosis. *Green Letters, 13*, 74–94.
Carter, P. (2010). *An auditorium for echoes: Inventing sound places and the ethics of recollection.* Paper presented at IASH Sawyer seminar 'Embodied Values: Bringing the Senses Back to the Environment', Old College University of Edinburgh, Edinburgh, UK.
Derrida, J. (1991). *Of spirit: Heidegger and the question* (G. Bennington & R. Bowlby, Trans.). Chicago: University of Chicago Press.
Derrida, J. (2002). The animal that therefore I am (more to follow). D. Wills (Trans.), *Critical Inquiry, 28*, 369–418.
Foer, J. S. (2002a). A primer for the punctuation of heart disease. *The New Yorker*, June 10.
Foer, J. S. (2002b). *Everything is illuminated: A novel.* London: Hamish Hamilton.
Foer, J. S. (2005) *Extremely loud and incredibly close.* London: Hamish Hamilton.
Foer, J. S. (2009). *Eating animals.* London: Hamish Hamilton.
Hanssen, B. (1998). *Walter Benjamin's other history: Of stones, animals, human beings and angels.* Berkley: University of California Press.
Harvey, D. (1996). *Justice, nature, and the geography of difference.* Cambridge, MA: Blackwell.
Heidegger, M. (1971). *Poetry, language, thought* (A. Hofstadter, Trans.). New York: Harper Row.
Heidegger, M. (1995). *The fundamental concepts of metaphysics: World, finitude, solitude* (W. McNeill & N. Walker, Trans.). Bloomington: Indiana University Press.
Holland, A. (2011). The value space of meaningful relations. In E. Brady & P. Phemister (Eds.), *Human-Environment relations: Transformative values in theory and practice.* Dordrecht, The Netherlands: Springer.
Ingold, T. (Ed.). (1988). *What is an animal?* London: Unwin Hyman.
Ingold, T. (2007). *Lines: A brief history.* London: Routledge.
Ingold, T. (2008a). *The environment as fluid space.* Paper presented at the 2nd Embodied Values workshop: Environmental aesthetics and ethics, IASH, University of Edinburgh, Edinburgh, UK.
Ingold, T. (2008b). *Feedback report* (Informal briefing to members of the 2nd Embodied Values Workshop). Edinburgh: IASH, University of Edinburgh.
James, W. (1890). *The principles of psychology* (Vol. 1). London: Macmillan and Company.

Kafka, F. (1954). *Beschreibung eines Kampfes: Novellen, Skizzen, Aphorismen aus dem Nachlass.* Frankfurt am Main: S. Fischer.

de Man, P. (1986). *Resistance to theory*. Manchester: Manchester University Press.

Rilke, R. M. (1989). *Duino Elegies* (S. Cohn, Trans.). Manchester: Carcanet.

Shaw, P. (2008). *Horses*. London: Continuum.

Shukin, N. (2009). *Animal capital: Rendering life in biopolitical times*. Minneapolis: University of Minnesota Press.

Spada, E. C. (1997). Amorphism, mechanomorphism, and anthropomorphism. In R. W. Mitchell, N. S. Thompson, & H. L. Miles (Eds.), *Anthropomorphism, anecdotes, and animals* (pp. 37–49). Albany: SUNY Press.

Wheeler, W. (2010). Gregory Bateson and Biosemiotics. *Green Letters, 13*, 35–54.

Chapter 7
The Intimacy of Art and Nature

David E. Cooper

Abstract This chapter critically explores a thought that has contributed to the perception of a sharp divide between human beings and their natural environments. This is the thought that there is a radical contrast between the aesthetic appreciation and significance of art and nature respectively. While attempts have been made to soften this contrast, some of these – both 'naturalistic' and 'constructivist' – implausibly reduce one type of appreciation to the other. A useful strategy for questioning the contrast is to consider 'hybrid' places, like gardens, which cannot be uniquely allocated to the category of art/artefact or to that of nature. Having rejected the idea that appreciation of gardens consists of two independent kinds of enjoyment – of art and of nature – the chapter argues that these hybrid places matter to us as symbols or epiphanies of a deep co-dependence between human beings' creative practice and their experience of nature. Traditions of creative practice shape experiences of nature, just as these in turn shape practice. Not even notionally is it possible to envisage a way of human life in which there is not this co-dependence. It follows from this that the alleged divide between the human and the natural receives no support from a proper understanding of the appreciation and significance of art and nature.

Keywords Art • Aesthetics • Gardens • Nature • Significance

Two aims of the research project from which the present volume developed were to question those perspectives that 'amplify the separation of humans from their environment' and to explore 'the interface of the human and the environment' as a 'value space'. One way in which, historically, the separation of human beings from their natural environments has been 'amplified' is through drawing a sharp distinction

D.E. Cooper (✉)
Department of Philosophy, Durham University, 50 Old Elvet, Durham, DH1 3HN, England, UK
e-mail: D.E.Cooper@dur.ac.uk

E. Brady and P. Phemister (eds.), *Human-Environment Relations: Transformative Values in Theory and Practice*, DOI 10.1007/978-94-007-2825-7_7,

between art and nature appreciation – or, better perhaps, between the significance that art and nature respectively have for us when we appreciate them. To query the sharpness of this distinction is, therefore, to question one aspect of an alleged separation of humans from their environment, and at the same time to consider one 'interface' of culture and nature as a 'space' into which evaluative, specifically aesthetic, concerns penetrate.

7.1 Appreciating Art and Nature: Difference and Dichotomy

A popular topic of refined conversation in the salons of eighteenth-century London, Edinburgh, Paris and Weimar was the respective merits of art and nature. Is nature such that it can and should be 'perfected' or 'improved' by art? Which is the more impressive phenomenon, a great mountain in all its sublimity or the great artist whose genius enables him to capture this quality in paint? These were not simply issues of 'Taste', of 'aesthetics' in a narrow sense of this term. At stake were the kinds of significance for human beings that art-works and natural places can possess. When in 1770 Arthur Young proclaimed that the artificial magnificence of Versailles was as nothing in comparison with 'the sportive play of nature in the vale of Keswick', he was not so much expressing an aesthetic preference as a judgement on the importance that art and nature should respectively have for us. Just the opposite judgement was expressed some years earlier by Charles Cotton when describing the gardens of Chatsworth as a 'paradise' in comparison with 'Nature's shames and ills' (quoted in Andrews 1999: 68, 71). It is no accident, as we shall see in the next section, that gardens figured prominently in the conversation on art and nature.

In our twenty first-century replacements for the salon – pubs, seminar rooms, blogs – the talk is no longer of perfecting nature or of the comparative merits of sublime nature and artistic genius. But the emergence of a widely accepted explanation of the accelerating degradation of natural environments has given a new pertinence and urgency to issues concerning the appreciation and significance of art and nature. It is by now something of a cliché of 'green' discourse that environmental degradation and devastation is grounded in a sharp opposition – the legacy, it is often charged, of Christian metaphysics – between the human and the non-human, between the realms of culture and nature. It is this dualism, one constantly hears, that has conspired to entrench a view of the natural world as simply a natural resource or storehouse of materials for our exploitation.

Enlightenment and Romantic contrasts between art and nature appreciation are not, of course, the only or even the main culprit accused of responsibility for 'amplifying' the larger opposition between the human and the natural. Still, making a stark contrast between the two forms of appreciation, and between the kinds of significance that art and nature respectively enjoy, has contributed to this amplification, and has influenced the terms in which it has been drawn. And if one is to understand, let alone endorse, the very general environmentalist ambition to dissolve the dualism of the human and the non-human, it is by questioning rather more tractable and particular dichotomies, like that between art and nature appreciation, where it would seem wise to begin. If there is an interface between these two, then this inspires confidence in a wider interface of the human and natural.

Questioning this dichotomy should not take the form of rejecting the several salient differences that exist between the ways in which art-works and natural things or places are typically experienced. Some of these differences are implicit in Theodor Adorno's (1997: 70) emphasis on the 'essential indeterminateness' that characterizes the experience of nature. This is an experience typically unconstrained by information about purpose or intention and by conventions, like those in force in the art gallery, that dictate the conditions of proper viewing. There is no reason, either, to reject the familiar point that appreciation of nature typically involves an active, engaged immersion and participation that has only minor parallels in the case of art appreciation. Nor should questioning the dichotomy weaken the principle – one too often ignored – that it is with the appreciation of art *as* art, and of nature *as* nature, that an understanding of the significance of art and nature must engage. (One does nothing to soften the contrast between two kinds of appreciation by holding that, for certain purposes, a natural object may be enjoyed as if it were a work of art – or vice-versa.) But to honour this principle, as well as the salient distinctions between the experiences of art and nature, is not to subscribe to a dichotomy of significance.

It is not, for example, to subscribe to G.W.F. Hegel's conviction that the discipline of aesthetics should be confined to the philosophy of art, precisely in order to 'exclude the beauty of nature' from its remit. This exclusion is justified, Hegel contends, because the beauty of art is of a different and 'higher' order: it is 'born of the spirit' in contrast with that of 'external spiritless nature' (Hegel 1975: 1–2). Art belongs to the sphere of freedom, creativity, spontaneity, spirit; natural things to the realm of necessity, causality, matter. The mountain stream or the swallow, therefore, must have a significance for us of an entirely different order from that of the sculpture or the sonata. Many later writers have followed Hegel's ruling that art appreciation is too remote from that of nature to be the subjects of a single coherent discipline, aesthetics.

These followers can, however, make strange bed-fellows. For art critics and aesthetes like Clive Bell, but also for many environmental ethicists, aesthetics should indeed exclude nature – but for very different reasons in the two cases. For Bell (1981: 20), it is the privilege and glory of art alone to communicate 'significant form', and for this reason 'aesthetics' – a term of honour – should be restricted to the study of art. For the environmentalists, however, the reason is that to regard nature aesthetically is to do so 'anthropocentrically', as a resource or storehouse, in effect, for the production of pleasurable sensations. (See, for example, Keekok Lee's 'ecocentric' criticisms of aesthetic attention to nature (Lee 1995)). In their own way, those who approach nature aesthetically are exploiting it for human benefit as much as those who treat it as a commodity.

If Hegel's followers form a mixed company, so do those whose ambition is to dissolve the Hegelian dichotomy by reducing one kind of appreciation to the other. On the one hand are 'naturalists', intent on demonstrating that both the production and experience of art are explicable in the same terms – those of natural selection – as our responses to the natural environment. The main contours, at any rate, of both art and nature appreciation are the ones established long ago when our distant ancestors' preferences turned out to have a survival value that those of other people – the ones who died out – did not. On the other hand are the 'constructivists', insistent that nature is simply our 'projection', a 'cultural artefact', so that in appreciating nature we are only enjoying our own handiwork. This suggestion is sometimes allied to an 'institutional' theory of nature appreciation corresponding to the well-known 'institutional' theory of art. In the cases of both art and nature, the idea goes, something acquires its status as an object of appreciation – as an art-work or as a scenic spot – in and through the decisions and practices of people (such as art gallery directors and tour operators) who control what the public will view (see Sepänmaa 2002).

There are, perhaps, elements of truth in these approaches, though each is rendered questionable by a sweepingly reductionist ambition. It is unclear, moreover, how they bear on the kinds of significance art-works and natural things have *for us*. Even if we were informed about the survival value for our distant ancestors of certain perceptions, or about the 'socially constructed' or institutionally managed character of the categories in terms of which we perceive things, could this have any place in our own understanding of why these perceptions matter for us?

7.2 Gardens and Other 'Hybrid Environments'

To explore this understanding, let's recall a staple ingredient of eighteenth-century salon conversations about the respective merits of art and nature: the topic of gardens, parks and other 'hybrid environments' that, as it were, are ambiguous between art and nature, and belong to a nature saliently affected by humanity. It is no accident that significant figures in the art-and-nature debate – Alexander Pope, Denis Diderot and Goethe, for instance – wrote at length about gardens. The garden, after all, was an eminently suitable peg on which to hang ruminations about the competing charms of art and nature. Rivalry between the respective aficionados of the informal 'English garden' and the formal 'French garden' condensed a wider dispute between the champions of nature, albeit 'improved', and those of artefact.

This was a rivalry important enough to surface in the writings of leading philosophers of the period, notably Immanuel Kant and Hegel. Kant approved of 'the English taste in gardens', holding that informal gardens obeyed his maxim that art should 'have the appearance of being nature' (Kant 1952: 88, 166). Hegel, who said he could not bear to look twice at 'English' or 'Chinese' gardens, responded that imitation of nature was disingenuous, and that, since 'man is the chief thing' in all artefacts, this is a thing that the gardens should proudly announce (Hegel 1975: 88, 166). Whatever one thinks of those rival claims, it is possible to agree that if one sets out, more generally, to reflect on how art appreciation and nature appreciation are related, it is entirely sensible to focus on items that appear to invite both.

One reason this is sensible is that, on the surface at least, gardens and other 'cultural landscapes', as the geographer Carl Sauer usefully dubbed them, constitute a problem for dichotomists. If artworks and natural things matter to us for quite opposed reasons, it is difficult to see how hybrid places could be appreciated at all. And, historically, dichotomists *have* had a problem with such places. Thus, for Hegel, gardening was an 'imperfect art', since gardens are, like monsters, neither one thing nor the other – neither art nor nature – and hence are 'worth nothing in themselves'. At best, they provide 'cheerful surroundings' for after-dinner strolls (Hegel 1975: 627, 700). For many of those environmental ethicists who we heard castigating the aesthetic appreciation of nature as 'anthropocentric', gardens are also a problem – for they are emblems, in effect, of human exploitation of the earth. It follows that if gardens and other cultural landscapes are *not* a problem – if, *pace* Hegel and the radical environmentalists, they are authentic, respectable and significant objects of appreciation – then it is with the dichotomy between art and nature appreciation that the real problem lies.

This is a conclusion which a dichotomist might resist, by arguing that appreciation of gardens is a combination of, or factors out into, two distinct kinds of appreciation – of art and nature respectively. But this is a bad response, for it is like arguing that the enjoyment of ballets factors out into two quite independent components – the enjoyment of music and the enjoyment of physical movement. In both cases, the suggestion is phenomenologically perverse, for typically, when we enjoy a garden or a ballet, we do not oscillate between quite distinct sorts of enjoyment that are then welded together. Our enjoyment and appreciation is altogether more holistic. (See Cooper 2005 and 2006: Ch 3.) This is why some philosophers – Crawford (1983) and Brady (2008), for example – speak of a 'dialectical relationship' between art and nature in the garden. The expression indicates that the significance of the whole depends on an 'interaction' or 'fusion' of artistic and natural aspects which results in an object of a 'third kind', one which is neither art nor nature. For example, Diderot judged that the garden at Marly succeeded because of a dialectical interplay and contrast between 'the delicacy of art in the bowers and the rudeness of nature in the dense bank of trees' (quoted in Adams 1991: 127).

Talk of a dialectical relationship here is reasonable enough, but it is not clear how it helps with our questioning of the alleged dichotomy between art and nature appreciation. For, granted that these can be 'fused' when directed towards 'third' objects like gardens, what can be gleaned from this about the appreciation of more 'pure' art-works, like paintings, and of wilder areas of nature that bear little or no human imprint? It is only if those 'third' objects were shown to be especially symbolic of something that is true of all things – of art-works and wild places included – that attention to them and to their significance could illuminate the relationship between art and nature appreciation. And if this is to be shown, the demonstration will need to extend well beyond talk of a dialectical interaction and fusion of art and nature in hybrid environments like gardens.

7.3 Co-dependence: Human Agency and Experience of Nature

The dialectical account, to its credit, recognized the striking co-dependence, in the garden or park, between human contributions and natural processes. It needs to be stressed that this co-dependence is not just of an obvious physical and causal kind, as when the rain damages the new fencing or the pruning affects next year's growth of roses. Equally important is co-dependence of an experiential or phenomenological kind. The placing of a statue may alter the perceived orientation of trees and plants, and the very formality of a garden design may render salient the roughness of the surrounding, uncultivated landscape. Conversely, the arrival of swallows might bring into perceptual prominence the old neglected shed in which they build their nest, while the 'borrowed scenery' beyond the garden wall may lend a mood to a garden that its maker could not otherwise have created.

Co-dependence of both these kinds makes the garden an especially apt symbol or epiphany of a deeper, more general co-dependence between human agency and the experience of nature. To gain a sense of this deeper co-dependence, we need to take the dialectical story beyond its exposure of the kind of interactions – obvious and less obvious – between the human and the natural just illustrated. For we can and should ask what it is that *makes possible* the interacting aspects – human purposes and creative agency, on the one hand, and the ways that the natural environment figures for us, on the other. These are not, after all, *givens*: for it is only in a context, or against a horizon or background, that people can have the purposes they do and that nature can show up for them as it does. In the final analysis, it is only in the context of a whole 'form of life' – of a whole way of 'being-in-the-world' – that people can have the purposes and experiences they do.

Crucially, the co-dependence between creative agency and experience of nature *goes all the way down*. For, in the total context of life that they presuppose there is, as William James (1977: 455) put it, no way of 'weeding out' and isolating the contributions of the human and the non-human. In particular instances, to be sure, we distinguish between these contributions. 'Did the gardener cut down the tree?' – 'No, the big storm blew it down'. But James's point is that there is no level of experience or description of the world, however simple and basic, which does not reflect human purposes and needs, and no level either which is unconstrained by the world's impact on our senses and imagination. And it is a co-dependence that goes all the way *back*, as well. For any creative engagement with nature is 'always already' informed by a way of experiencing nature, just as the way nature presents itself to us is 'always already' shaped by our purposive engagement with it. This is Martin Heidegger's point in the celebrated passage on a Greek temple in his essay on 'The Origin of the Work of Art' (1971: 41ff). The temple – or, more generally, Greek art and craft – gave to natural things, like eagles and rocks, their 'look', and enabled 'earth' to emerge for the Greeks as it did. But, at the same time, the 'look' of things and the emergence of 'earth' helped to shape the Greeks' 'outlook on themselves', an outlook that in turn helps to shape their culture, politics and creative ambitions.

It is all too easy, of course, to take for granted not only the purposes and practices familiar in our own culture but also our prevailing perceptions of the natural order. And when these are taken for granted and remain inconspicuous, the rich context of co-dependence between purposes and perceptions is forgotten. That is why, perhaps, Heidegger chose an ancient culture very different from our own to illustrate his point, and why we too, if we are properly to appreciate deep co-dependence, do well to reflect on practices and perceptions unfamiliar or alien to us. To reflect, for instance, on the design of the Japanese tea house garden, or on aboriginal rock art. Reflection, in both cases, soon renders conspicuous the extent of co-dependence in the context that makes such practices possible. For these are practices that presuppose ways in which nature shows up for people, but these ways themselves owe their possibility to just such practices of design and depiction.

Once the extent and depth of co-dependence are recognized, it is by reference to this co-dependence that it is plausible, even compelling, to understand the immense significance that gardens have for many people. When philosophers speak of gardens 'mediating' between or 'reconciling' art and nature, they are gesturing towards the thought that gardens are epiphanies of a co-dependence – one that goes all the way down and back – between creative human agency and the reception of nature.

This thought is an important one but, as it stands, it does not address the issue of the appreciation of things that are not 'hybrid' – things which include, of course, 'pure' art-works and wild places. How could reflection on the significance of gardens contribute to an understanding of the kinds of significance that a painting or an iceberg have for us? Such worries, though, betray a misconception about the reason for focusing on gardens and other 'hybrids'. The point was not that these objects of a 'third kind' are at all unique in exhibiting a co-dependence of human agency and experience of nature. The point, rather, was that they are – in virtue of the very conspicuous ways in which they combine art and nature – especially striking epiphanies or symbols of this co-dependence. The thesis of co-dependence, towards which we were led by reflection on these epiphanies, is totally general. The sense of co-dependence, though initially inspired by attention to gardens and the like, cannot remain restricted to them alone.

It is a sense that will smoothly extend, for a start, to what are sometimes called the 'middle landscapes' that stand between patently 'humanised' sites, like gardens, and genuinely wild places. These landscapes correspond roughly to what, in Britain at least, we have in mind by 'The Countryside', with its grazed hillsides, coppiced woods, and pastures. Here, we can hardly fail to recognize, as R.G. Collingwood did in 1925, the 'spectacle of a rural society living in the pursuit of traditional arts and deeply rooted in a landscape which has in part created it and in part been created by it' (quoted by Benson 2008: 153). Once the sense of co-dependence has extended in this way, it will eventually extend to all things, for we come to reflect that there are no human achievements, those of painting included, that are without dependence on ways of receiving nature. Indeed, if Merleau-Ponty (1964: 16) is right, it can be the purpose of some painters to expose precisely this dependence. The sensitive viewer of Cézanne's paintings, he remarks, will recognize how they 'reveal the basis of inhuman nature upon which man has installed himself'. Equally, we come to reflect that no experiences of natural places, however wild, are without dependence on human projects and purposes. Modern paintings and virgin forests may not, like gardens and to a lesser extent The Countryside, wear their dependences on the sleeve, but it is only in the context of a general form of life, in which human agency and an experience of nature are always already mutually shaped and shaping, that people can paint in a certain style and be attuned in a certain way to natural environments. As the primatologist, Frans de Waal (2001: 170f), puts it, there is a 'dual inheritance at work … there's plenty of nature in culture, just as there is plenty of culture in nature'.

7.4 Art, Nature, Significance and Life

The significance of our encounters with art, like that of our encounters with nature, is ultimately understood by reference to the total context of life in which they are possible. This does not entail, of course, that they signify the same, for it is, typically, along different reflective routes that we travel when seeking to understand the significance of art-works and natural things respectively. In the case of an art-work, typically, the route is through Art – through the history, traditions and trajectories of a culture to which the work belongs. The route, to recall Heidegger's distinction, is through what has directly shaped and mirrored our 'outlook on ourselves'. In the case of a natural thing, the reflective route typically takes us through Nature – through the history, processes, and cycles of a system in which it has its place. Here the route is, as it were, through what gives things 'their look'. But, as already observed, our outlook on ourselves and the look of natural things are not finally separable, and it is important that, while honouring differences in our experience with art and nature respectively, these are not dichotomised. The histories of people's outlooks and of the looks of natural things could not be what they are except in their mutual intimacy.

There is much to learn here from that neglected scientist and phenomenologist of the imagination, Gaston Bachelard. The main theme of his later writings, such as *The Poetics of Space*, is *reverie* – the name, for him, not of idle day-dreaming, but of a fluent, open, unencumbered, spontaneous style of reflection that is invited or prompted by the words, images and things that matter most to us. A lesson we learn from him is that it makes less difference than one might suppose to the routes of reflection a person follows where, as it were, he or she starts – with an artistic image or with a natural item. Many of Bachelard's own reveries are occasioned by lines of poems, but many, too, are responses to experiences of items such as bird nests or flowers. 'A nest', he writes, 'is a precarious thing, and yet it sets us to a reverie of security', just as a poet's image of a nest 'repulsing death' may prompt reflection on 'cosmic confidence' (1994: 102–3). Whatever the occasion might be, the subsequent reverie will, fluently and flexibly, invoke the products of both artistry and natural process. It will, for example, glide from poetic images to the things they are images of, and back again.

This implies, of course, that the two routes of reflection, through Art and through Nature, are ones that cross at many points and, indeed, that for whole stretches they share the same carriageway. In my own case, for example, I can imagine no reverie on why certain hills, like the Cheviots in the North-East of England, matter to me that would not, at some junction, invoke the paintings of Turner. Nor, conversely, can I imagine a reflection on why this particular painter is important to me that would not evoke thoughts about natural places of the kind he painted. The ease with which our eighteenth-century forebears switched between reveries on artworks and nature is indicated by their enthusiasm for the Picturesque – a term that came to be applied both to paintings, like those by Salvator Rosa, and to the kind of rugged, rustic places they depicted. Essentially to the appreciation of scenery as Picturesque was a readiness to recall paintings of a corresponding mood.

Bachelard reminds us, too, that however diverse the occasions for reverie may be, and however distinct the routes our reveries may take, the destination – or, as he calls it, the 'resonance' – is always the same. For what reveries yield is 'the whole of the whole … a unity of world', which is compared to a 'gigantic fruit' whose segmentation into art and nature, into human and non-human portions, is the work of abstraction (Bachelard 1971: 175). In my more prosaic terms, the routes of reflection converge on a form of life, a way of being-in-the-world, where art and nature are deeply, inseparably co-dependent.

Bachelard's *rêveur* is someone for whom 'the world no longer poses any opposition', whose 'I no longer opposes itself to the world': he or she is, in effect, someone who no longer distinguishes subject from object (Bachelard 1971: 167). Perhaps he is right: perhaps this dissolution is the final accomplishment or 'resonance' of reverie. But this conclusion of Bachelard's goes beyond what I have proposed in this essay. My more modest proposal has been that, whatever the eventual fate of such dualisms as those between subject and object, the human and the non-human, the spheres of freedom and necessity, or of spirit and matter, they cannot expect to be fortified by an alleged dichotomy between the significance that art and nature respectively have for us. For reflection or reverie on their significance invites, demands even, a sense of their co-dependence, one manifested in, but not confined to, such 'ambiguous', 'hybrid' places as gardens. This is a sense that cannot easily be combined with dualistic visions.

To put the matter in terms of the issues that prompted this chapter, I have argued that at least one of the ways in which 'the separation of humans from their environments' has been 'amplified' is illegitimate. Reflection on the differences, real as they are, between the appreciation and significance for us of art and nature respectively, does not endorse the idea that, in effect, we are confronted with two different worlds, a cultural and a natural one. How extensive the 'interface of the human and the natural environment' might be is a large question: but if I am right, there exists an interface in the form of a 'value space', for the modes of appreciation and admiration invited by art and by nature are ultimately inseparable.

References

Adams, W. H. (1991). *Gardens through history: Nature perfected.* New York: Abbeville.
Adorno, T. (1997). *Aesthetic theory* (R. Hullet-Kentor, Trans.). London: Continuum.
Andrews, M. (1999). *Landscape and modern art.* Oxford: Oxford University Press.
Bachelard, G. (1971). *The poetics of Reverie: Childhood, language, and the cosmos* (D. Russell, Trans.). Boston: Beacon.
Bachelard, G. (1994). *The poetics of space* (M. Jolas, Trans.). Boston: Beacon.
Bell, C. (1981). *Art.* New York: Chatto & Windus.
Benson, J. (2008). Aesthetic and other values in the rural landscape. In S. Arntzen & E. Brady (Eds.), *Humans in the land: The ethics and aesthetics of the cultural landscape* (pp. 141–57). Oslo: Unipub.
Brady, E. (2008). Relating humans and nature through agricultural landscapes. In S. Arntzen & E. Brady (Eds.), *Humans in the land: The ethics and aesthetics of the cultural landscape* (pp. 121–40). Oslo: Unipub.
Cooper, D. E. (2005). Gardens, art, nature. In N. Kingsbury & T. Richardson (Eds.), *Vista: The culture and politics of gardens* (pp. 5–12). London: Frances Lincoln.
Cooper, D. E. (2006). *A philosophy of gardens.* Oxford: Clarendon Press.
Crawford, D. (1983). Nature and art: Some dialectical relationships. *Journal of Aesthetics and Art Criticism, 42*, 49–58.
de Waal, F. B. M. (2001). *The ape and the sushi master.* London: Penguin.
Hegel, G. W. F. (1975). *Aesthetics* (T. Knox, Trans.). Oxford: Clarendon Press.
Heidegger, M. (1971). *Poetry, language, thought* (A. Hofstadter, Trans.). New York: Harper & Row.
James, W. (1977). Pragmatism and humanism. In J. McDermott (Ed.), *The writings of William James* (pp. 449–461). Chicago: University of Chicago Press.
Kant, I. (1952). *The critique of judgement* (J. Meredith, Trans.). Oxford: Clarendon Press.
Lee, K. (1995). Beauty for ever? *Environmental Values, 4*, 213–226.
Merleau-Ponty, M. (1964). *Sense and non-sense* (H. L. Dreyfus & P. A. Dreyfus, Trans.). Evanston: Northwestern University Press.
Sepänmaa, Y. (2002). The two aesthetic cultures: The great analogy of art and the environment. In A. Berleant (Ed.), *Environment and the arts: Perspectives on environmental aesthetics* (pp. 39–46). Aldershot: Ashgate.

Part II
Transformative Values in Practice

Chapter 8
Embodying Climate Change: Renarrating Energy Through the Senses and the Spirit

Michael S. Northcott

Abstract The refusal of liberal capitalist societies effectively to regulate greenhouse gas emissions reflects the underlying Cartesian assumptions that persist in the culture of these societies. Merleau-Ponty, Karl Bohm and others have sought since the 1950s to unsettle the Cartesian legacy by recovering a more embodied and sensual account of human cognition and consciousness. This philosophical repair finds significant echoes in the narrativist turn in theology, and both recall approaches to knowing and action that may be found in the theology of Saint Paul, and in traditional Christian ritualisation of relationships between past, present and future action and being. The enduring and embodied character of such inter-temporal beliefs and rituals represents a sensitivity to inter-generational relations concerning energy use and climate change which is needful if modern perceptions about the consequences of energy use for the planetary future are to change.

Keywords Climate change • Energy • Human • Future • Body

Liberalism as summarised by John Rawls in his *Theory of Justice* and *Political Liberalism* has one defining characteristic which may be traced back through Adam Smith, John Locke and Thomas Hobbes and this is the division of labour that Hobbes, Locke and Smith erected, and that Rawls sustains, between the individual and the body politic (Murphy 1998). From Hobbes to Rawls the individual is described as an autonomous creature of her own making who is guided by desires and projects which are the outcomes of her own self-construction. The role of the body politic is not so much to create the conditions for virtuous individuals to develop and flourish as to impose minimal conditions of order, such that for example property rights are respected, on the differing desires and sometimes competing

M.S. Northcott (✉)
School of Divinity, New College, Mound Place, Edinburgh EH1 2LX, Scotland, UK
e-mail: M.Northcott@ed.ac.uk

E. Brady and P. Phemister (eds.), *Human-Environment Relations: Transformative Values in Theory and Practice*, DOI 10.1007/978-94-007-2825-7_8,

projects that individuals construct. In effect liberalism conceives of desire in terms of a market analogy. The function of the body politic is to sustain a realm of relatively unconstrained interaction where minimal conditions for civil peace and social order are maintained. The modern condition of liberty is one in which individuals are not constrained in their pursuit of their desires by the needs of others who are less successful than they in pursuing and achieving their desires and projects: the market will act to direct individual desires towards a collective state in which all achieve greater utility. The body politic is left with the task of siphoning off a portion of market goods to enable the construction of public institutions in which property rights are respected, and where necessary defended, and the minimal conditions of bodily life are granted to those who are less successful in pursuing their own projects of self construction. This division of labour between individual desiring agents and the body politic means that for liberalism, provided the body politic functions minimally to sustain the conditions for respect of individuals and property, individuals do not themselves have to pursue justice in their own projects; they can defer the moral requirement of the pursuit of the common good to the invisible hand of the market, and to the public institutions that, via taxation, their activities fund and sustain.

This division of labour between individuals and the body politic in liberal political theory, and in the economic and political arrangements of the modern nation state and the market, was first decisively criticised by Leo Strauss in the 1960s in Chicago. For Strauss the problem with liberalism was the idea that it was possible to construct a good society through institutions and procedures while not requiring the individuals within it to be good. Instead of this Strauss proposed a revised Aristotelianism in which intelligent and wealthy individuals need again to be instructed in the virtues so that they can act out of their own superior accomplishments and largesse to better the condition of the poor with the beneficial corollary that both elites and masses might then be freed from the undue constraints, and moral hazards of an over-weaning and interventionist nation state (Strauss 1995).

Strauss's neo-Aristotelian critique has been highly influential among Anglo-American politicians. There are clear echoes of Strauss in the rhetoric of compassionate conservatism first mobilised by Margaret Thatcher in Britain in the 1970s, and again by George W. Bush since 2002 (Northcott 2004a). Both advanced tax-reduction and avoidance strategies for the rich – reducing the numbers of tax inspectors as well as cutting wealth taxes – while reducing welfare for the poorest on the basis that compassionate individuals and the charities and churches they support are putatively better able to respond to the needs of their poor neighbours than the bureaucrats or public servants of the nation state. Added to this account of the compassion of the wealthy, famously advanced by Margaret Thatcher in her exegesis of the parable of the Good Samaritan (Northcott 2004b), is a critique of welfare and the nation state whose largesse to the less fortunate is said to be morally hazardous because it rewards laziness and other vices among the poor while at the same time, because of the taxation required to sustain it, undermining the motivation of the wealthy to create wealth.

The critique of the nation state as morally hazardous when it tries to embody morality in institutions that promote equity and justice has been taken up not just by political leaders from the right but by economists and others who advocate economic globalization. In this project the nation state is said to take a back seat while economic actors under the new libertarian conditions of international trade, guided by international institutions such as the World Trade Organisation, become principal drivers of increases in human welfare. Against the consistent critique of the globalisation project from civil society groups, feminists, philosophers and theologians, publicly embodied in the form of the 'Make Poverty History' campaign around the G8 meeting in Scotland in June 2005, the WTO, the European Commission, and the American Federal Government, as well as economic corporations and most corporately-owned media, continue to press on wealthy and poor countries alike the instrument of borderless international trade as the principal means of achieving progress in prosperity and welfare for rich and poor alike in the twenty-first century. In the globalization project the division of labour between individual actors and the body politic reaches a new extreme of alienation where the principal function of the body politic is no longer to ameliorate market failures so much as to promote the invasiveness of the market even into those areas of human life where in the past public institutions have been constructed to resolve market failures. This is what the Private Finance Initiative, sometimes called 'Public Private Partnership', represents in Britain. Its international form is the larger project of privatisation and trade in services under which the intention is that everything from education and health care to energy and water utilities are ultimately run on the market model of private provision and user chargers.

Given the extreme libertarian outcomes of the Straussian critique of liberalism it would appear that this critique was in some respects crucially flawed even on its own elitist terms. There is little evidence that the new more extreme conditions of liberty for capital and property owners have actually produced a social condition in which individual morality has clearly advanced in terms of its impact in ameliorating social ills. On the contrary both Britain and North America, the two societies which more than any other have embraced the Straussian critique of liberal social contract approaches, have suffered dramatic increases in inequality and violent crime in the last thirty years. Many town and city centres in England in the present decade are characterised by mass drunkenness and disorder at night, while in the United States many inner city areas are no-go areas even for the police at certain times of night. Far from liberating individuals to behave more virtuously, the increasing pursuit of market solutions to social ills has produced a situation in which millions are deprived of civil peace and social order on their own streets.

In the midst of this ideological and political condition of late liberal capitalism emerges the spectre of anthropogenic climate change. Sea level rises of 20 cm are the most visible sign of the effects of global warming but predicted sea level rises from polar ice melt range from 1 to 7 m. Changes in land and ocean temperatures are producing dramatic effects in terms of the movement of species, and the spread of deserts. In 2005 locusts were for the first time in recorded history breeding in Southern France, while temperatures in the heavily populated state of Orissa in the

Indian sub-continent now regularly exceed 40°C. Climate change challenges the foundations of liberal political economy, and in particular the liberal division of labour between the amoral desiring individual and social institutions which embody and promote the common good, because it suggests that under conditions where individual consumers and corporations maximise their preferences with the minimum of moral constraint the long-term health and stability of the planet and all its inhabitants are threatened. It should not surprise us then that Straussian-influenced globalizers, including American Senators and Presidents as well as Australian and British politicians, have opposed government-led efforts to conserve energy and have for many years denied or ignored the scientific evidence for climate change: climate change more than any other modern phenomenon represents a radical challenge to political liberalism, and to its neo-liberal recasting in the guise of the 'free' market, globalization, and the minimalist state.

The collective pursuit of the project of economic growth through unfettered consumption has been advanced on the basis of the release of seemingly limitless quantities of energy from the earth's crust, first in the form of coal to fuel the earliest steam-driven machines of the industrial revolution, and latterly oil to fuel internal combustion engines, electricity generators and jet engines. These fossil fuels represent the prehistoric warmth of the sun laid down as carboniferous biomass in the earth's crust as plants and sea creatures turned this energy into oxygen and carbon in the course of geological time. Until the discovery of anthropogenic climate change there appeared to be no biophysical limits to the amount of stored energy that could be released into the earth's atmosphere and hence to the size of the energy-driven human economy. However, with a current net annual output of seven billion tons of carbon per annum into the atmosphere, the modern human economy is seriously exceeding the capacity of atmosphere, forests, oceans and soils to absorb its energy emissions. The oceans are already replete with the excess carbon output of the industrial era which they have taken up in the last 100 years and as they are unable to absorb CO^2 at the same rate, fossil fuel emissions now increasingly end up in the upper atmosphere, enhancing the greenhouse effect and driving up oceanic and air temperatures and thus fuelling more extreme weather events and ice melts of a kind never before experienced in the 15,000 year span of the present Holocene era. Climate extremes were common before the Holocene era and it was precisely the new stability of CO^2 levels, and hence of relatively stable land and ocean temperatures, which enabled the development of human agriculture and cities, and the dramatic expansion of human numbers, in the present geological era (Northcott 2007).

Climate change represents a challenge not only to energy-led consumerism and unfettered capitalism, and its latest guise in the form of borderless global trade, but to the epistemological and ontological foundations of modern liberalism. At the heart of Rawlsian liberalism, and its neoliberal offshoots, lies the assumption that individual actors are seats of consciousness, desire and rational decision-making who are intrinsically autonomous from other bodies and from the biophysical environment. It is this assumption which explains the liberal division of labour between individual agency and the body politic; political institutions embody morality in the relational world of public space but individuals are conceived as essentially independent of this bodily domain, their actions determined by their inner desires and rational choices rather than by their biological relations to other agents and to the environment.

Criticism of this essentially disembodied conception of agency and consciousness, which is rooted in the Scottish and German Enlightenments, has come from various quarters and nowhere more powerfully than in the phenomenological critique of Maurice Merleau-Ponty. A key element in Merleau-Ponty's ontological move is his attempt to re-inscribe community and time upon the acts and gestures of individual bodies:

> What is true, however, is that our open and personal existence rests on an initial foundation of an acquired and stabilized existence. But it could not be otherwise if we *are* temporality, since the dialectic of acquisition and future is what constitutes time. (Merleau-Ponty 1962: 384)

In this account the perception and effect of every individual action or gesture already involves an unconscious but real projection of the self into relations with other bodies in the past and the future in such a way that they are taken up into a larger embodied pattern which Merleau-Ponty characterised as 'existential rhythm' (Zaner 1971). Also in the French philosophical tradition, Marcell Mauss proposed in his essay 'Techniques of the Body' that the body is the primary tool through which humans give shape to the biophysical world, and that it is at the same time from the body that the human world – or what I above called public space – is also constituted, hence the very term 'body politic' (Mauss 1973). Mauss and Merleau-Ponty directed anthropologists and philosophers to a reconsideration of the place of the body in human cultural and social construction as well as in human consciousness and their insights have been taken up in ecological thought, psychology, philosophy and the natural sciences. Thus quantum physicist David Bohm posited what he called an 'implicate order' in the substructure of atomic life which connects individual acts and the biophysical structure of the world (Bohm 1980), while Gregory Bateson constructed an ecological account of consciousness (Bateson 1973), and John Gibson proposed an ecological account of visual perception (Gibson 1979). For Bateson and Gibson human consciousness and perception are constituted by their location in the body and by relations with other bodies, human and nonhuman. This recognition involves the rejection of reductionist and atomistic accounts of rationality which have achieved widespread currency under the influence of René Descartes and Immanuel Kant, and which reach their nadir in economic liberalism. Advocates of a more embodied approach to epistemology and perception argue that both philosophers failed to take account of the multi-sensory and ecologically-situated character of human being (Burkitt 1999).

The recovery by phenomenologists and natural scientists of the embodied character of human identity and perception also finds echoes in the turn among some modern moral philosophers and theologians to narrative, story and tradition as key elements in human development and rationality (Hauerwas and Burrell 1989). The life projects, desires and decisions of individuals in a narrativist perspective cannot be understand apart from knowledge of the life-story, and cultural memes, that have shaped an individual in her development, including relations with significant others, and with particular bioregions, communities, places, practices and traditions.

In this perspective climate change is the exposure of this deep biophysical and storied structure underlying individual agency in the biogeochemistry of the earth and of the mutually constitutive relations between animals, atmosphere, oceans, plants and rocks that was first theorised by James Lovelock in his 'gaia hypothesis'. Interactions between organic life and planetary biogeochemistry memorialise the record of past actions, particularly in the industrial era, in present and future time. Thus the narrative of climate change begins not with the discovery at the Mauna Loa observatory in Hawaii of raised atmospheric levels of CO^2 in the 1970s, nor even with Svante Arhenius' original proposal in 1896 that industrial carbon could enhance the greenhouse effect (Arhenius 1896), but with the eighteenth century invention of the steam engine by Thomas Newcomen, used as it was to facilitate the mining of coal at the origins of the English industrial revolution (Landes 1969)

Advocates of a phenomenological approach to human being and knowing have for more than fifty years attempted to unseat the regnant constructivist and rationalist assumptions in Western political and philosophical thought which underlie the project of liberalism. However, they have been largely unsuccessful in dethroning the cultural power of these assumptions given the strength of the modern commitment to the ethereal economy of unfettered desire and the 'free' market, and the relative inattention to the biophysical economy of organic bodies in the earth system (Dupuy 2004). It may be that climate change offers not only a decisive moment in the earth's history, but a radical challenge to the transformation of the earth that modern industrialism has presaged where the rhythm of the earth and the bodies, both individual and collective, of humans are brought into a new alignment.

Attempts to mitigate climate change have so far failed to evince or manifest the kind of radical rethinking of the liberal project that an embodied approach to climate change would suggest. There are many reasons for this, not least the continuing influence of the disembodied account of desire and individual agency sustained by the majority of modern Western philosophers. Equally important is that climate change mitigation depends upon collective action by a whole host of actors including consumers, corporations, governments and international agencies. For the behaviours of such a diverse range of actors to be directed towards the shared goal of reducing fossil fuel emissions, so stemming the future consequences of climate change, requires a degree of coordination and cooperation which would at first hand seem hard to achieve, although the current neoliberal economic project of global borderless trade in goods and services does represent just such a form of global cooperation and coordination. But this neoliberal project is in direct opposition to the goal of limiting global carbon emissions. When Africans are encouraged by the current regime of world trade to grow mangoes for export to Northern Europe, and while American and European farmers use government subsidies to purchase energy-intensive fertilisers and farm machinery so they can export wheat to Africa, then the industrial food economy is given over to a model of carbon consumption which is clearly a major progenitor of rising carbon emissions.

Given the epistemological and ecological inadequacies of the liberal and neoliberal narratives of (disembodied) private rational choice and (embodied) public morality it is unsurprising that institutional procedures influenced by this narrative have produced an international climate change treaty process which is ineffective in promoting real carbon reductions. But the other principal root of the problem is the idealistic character of the cost-benefit calculations which economists apply to the problems of either adapting to or mitigating climate change. Bjorn Lomborg articulates a widespread bias amongst economists and industrialists when he suggests that the costs of mitigating the future effects of climate change are so great, and the nature of these effects so uncertain, that it is more economically beneficial to plan to adapt human behaviours and procedures to climate change when it occurs than to regulate economic activity so as to reduce present carbon emissions so that these potential future costs may be reduced (Lomborg 2001). This argument is predicated on the economic practice of social discounting which compares present and future economic activity and, on the basis of current and predicted interest rates – and hence the growth in value of money saved – argues that future activities cost less than present ones.

The problem with such economistic dismissals of the need to mitigate climate change in the present is not just that they undervalue costs to future generations of climate change but that they involve measurements of cost and benefit which are so theoretical as to misrepresent the real world of biogeochemistry. The reason is simple: cost benefit calculations are conducted on the basis of theoretical economic rules of supply and demand, and monetary accounts of profit and loss, both at corporate and at national level. However, these rules and accounts notoriously fail to count as costs many of the environmental and social costs, dubbed 'externalities' by economists, which economic activities impose upon individual and collective bodies in the real world (Spash 2002). Climate change is the most dramatic and long-lasting of all such 'externalities' but as yet neither corporate nor national accounting systems include climate change effects in monetary measures of economic activities. Again we are dealing here with the problematic outcomes of the disembodied character of modern Western accounts of rationality and rational choice behaviours, and in particular their influence on the way in which economists model, and accountants measure, the behaviours of individuals, firms and governments.

There is a further feature to the problem of disembodiment in the procedures by which liberal and neoliberal economists and politicians respond to signals from the biophysical environment, and in particular climate change. It concerns the thorny philosophical problem of time. A fully time sensitive narrative of climate change would start not at the invention of Newcomen's steam engine but way back in prehistory for it would involve the narration of the way in which myriad creatures in the oceans and on *terra firma* absorbed the prehistoric warmth of the sun through photosynthesis and eventually drew down the resulting carbon biomass in the form of fossil fuel deposits in the earth's crust or under the ocean floor (Sieferle 2001). The extraordinary growth of the industrial economy in the last 150 years effectively rests upon the sudden release of this geological record of millions of years of solar warmth into the planet's present biogeochemistry in the form of carbon emissions.

In other words the great range of goods and services that the modern consumer is encouraged and stimulated by a vast advertising industry to enjoy and expend rests upon a hidden history of organic life which is hundreds of millions of years old. Yet this biophysical exchange of organic heritage for present consumption takes place in a cultural context where there is an increasingly attenuated sense of the connection between past, present and future generations, practices, stories, and traditions. Again the disembodied conception of the desiring and choosing rational self is implicated here for, as Edmund Burke argued, the desire and willingness to *conserve* a state of affairs for future generations rests upon a consciousness of connection between present and past which the revolutions that inaugurated the modern age, explicitly rejected; 'people will not look towards posterity when they fail to reflect on past generations' (Burke 1790: 141).

The strange timelessness which informs present cultural consciousness may be said to originate in philosophical problems with time in theology and philosophy which precede the birth of the modern era. The origin of these problems was in the Scholastic notion of occasionalism in which late medieval theologians – such as Ockham and Duns Scotus – posited that an omnipotent God could act in the world so as to disturb the laws of nature and thus that the physical appearance of order in nature, as present humans observe it, may be illusory. For the nominalists there was an unfathomable gap between the purposes of God in creating the world, and in continuing to intervene in that world, and the physical appearances of things. Hence each individual instance of being bore no settled relation to the first cause and eternal origin of being. This produced a capricious instability to physical appearances, and cause-effect relations such that divine purposes could no longer be read off from the material creation (Dupré 1995). This move in Christian metaphysics was taken up by seventeenth century occasionalists such as Malebranche, who, following Descartes, argued against there being necessary connections either between bodies or between mind and body (Nadler 2011; Phemister 2011). This represents a real rupture in the relations between inner consciousness and outward appearances, and between past, present and future in Western ontology, and ultimately Western culture, that efforts to retrieve embodiment and narrative in philosophy and theology are only now beginning to address.

The narrative theologian addresses this problem by arguing for continuity in the relations between past and present such that the present possibility of Christian existence issues from the prior calling of Abraham, Moses, and above all the calling of Christ himself, upon which Christian identity, practices and traditions ultimately rely. So for example when the Apostle Paul claimed that Jesus Christ is the same 'yesterday, today and forever', he was arguing for continuity in this new religious tradition with prechristian and future time, and this continuity in Christian history takes a number of institutional forms: these include practices such as the Christian sacraments of baptism and Eucharist, the public reading of the scriptures in which the stories of God's ways with God's people are recorded and rehearsed, and the recording and remembering of the lives of the saints. In a narrativist perspective what connects present Christian actions with past inheritance are these shared practices and stories, as well as a shared spiritual relation, expressed in private prayer as well as public worship, to God. How though might this sense of cultural and spiritual continuity be said to relate to the problem of disconnection between past, present and future in relation to energy consumption and climate change?

One approach is to argue for a new consciousness of the moral claims of future people on presently existing people, a claim which would involve a redefinition of the command of Christ to 'love your neighbour as yourself' in which the category 'neighbour' includes future persons who, though distant from present persons in time, are nonetheless connected to them by the planet's biogeochemical memory which visits the effects of present actions on future people (Northcott 2010). This idea of connection between past, present and future actions and generations is also encapsulated in the Christian conception of original sin, most fully elaborated by Augustine in the fourth century, and with antecedents in St Paul, and in Jewish reflection on the 'sins of the fathers' which the Torah represents as being visited on 'the third and fourth generations'. The Christian claim to find in the Christ events liberation from the inheritance of the ill effects of the actions of previous generations is that Christ displayed in his incarnation, dying and rising a new form of creaturely being – a 'new creation' – in which the body is at last liberated from its captivity to sin and death which are represented by St Paul as the inheritance of the actions of previous generations from Adam and Eve until the birth of Christ. This liberation takes the bodily and social form of new community, the 'body of Christ' in which individuals are joined in a common quest for peace and reconciliation in which past sins are forgiven and just relations between rich and poor, weak and strong, are pursued and sustained.

However Lynn White and others argue that, far from advancing a conservationist ethic with respect to the earth this religion of liberty desacralised the biophysical world so that the mortal world became amenable to destructive re-engineering by humans in their pursuit of their new freedoms (White 1967). Thus Christian merchants who employed forced labourers to mine coal from under the skin of the earth that fuelled the industrial revolution had already overcome any fears they might have had that such an act would be irreverent to mother earth (Merchant 1989). For White, the Christian story of liberation from the law of sin and death means that Christians attend less to the biophysical consequences of their actions than those of other traditions – such as animists or Hindus – who continue to believe in something like earthly Karma or 'what goes around comes around'.

These arguments have been played out extensively in the growth of ecological theologies in the last forty years. What I am claiming here is that the problems I have identified with respect to embodiment, the ethereal economy, and emissions from energy consumption, require some rethinking of our Western cultural traditions around what constitutes selfhood, rationality, and public morality. In particular I am arguing that the problem of climate change requires a renewed appreciation of the narrative relations between past, present and future generations, and between past, present and future ages of the earth, of the kind sustained in the Christian tradition through such practices as scripture reading and remembering the lives of the saints. I am also suggesting that these narrative relations take a distinctive political form in the idea of the 'body of Christ' which is a political metaphor for the relationships between rich and poor, weak and strong, who belong to particular communities of Christians in different parts of the world and who are united in their communion by their common devotion to the sovereignty of Christ as the 'head of the body'. This idea of the 'body of Christ' carries with it for St Paul a clear moral implication that Christians in one part of the world have obligations to Christians in other parts of the world who may be suffering from famine or some other kind of need. As St Paul suggests to the Christians at Corinth, their abundance finds its correlate in the need of Christians in Judea who, after the Jewish rebellion and its vicious putting down by the Romans, were suffering from famine (2 Corinthians 8).

How though might this ancient and idealised picture of the body of Christ interact with the modern problem of international relations, and intergenerational relations, presented by climate change? Clearly the implication is that the primary generators of climate change – Western consumers and corporations – have moral duties to the people who are already experiencing the effects of climate change in other parts of the world, and to future persons who will experience these effects in ever larger numbers. But the energy economy, with its electricity generators, connective grids and pipelines, petrol pumps, and electric switches, distances energy users from the effects of their energy consumption. And hence the title of this paper; the urgent need is to recover ways in which households and corporations can reconstitute their daily embodied rituals with regard to energy use so as to reconnect them with their source in the creation story from the beginning of time, and hence with posterity.

Celtic Christians had a particular set of practices with regard to energy use which contained something of what I am getting at. When they went out to dig peat, and when they laid it on the fire each morning to warm the household and cook its food, they undertook these actions in the sacred name of the Holy Trinity, accompanying these actions with prayer and reverence to the creator God. They saw the fire, which they would make from three pieces of peat, as in some sense holy, as a physical analogy of the Sacred Trinity, and as a manifestation of the fire of the Holy Spirit, who is the immanent presence of the Trinity and who warmed their hearts in worship even as the hearth warmed the household. Medieval architects had a similar reverence for the light of the sun which they expressed in the design of the great European cathedrals which were constructed in such a way as to draw down the light of the sun onto the floor of the chancel, nave and sanctuary through the great high windows which the new gothic building techniques enabled them to construct. Similarly William Blake in the nineteenth century called the energy which all around him was being released and mobilised in the industrial revolution 'divine energy' and so reminded his readers that this energy was not purely physical; it originated in the divine light which first shone on a dark and formless universe to bring forth life from the oceans. For Blake this recognition of the divinity of energy did not justify its release. On the contrary it was central to his apocalyptic critique of coal-fuelled industrial England in which workers were reduced to denizens in the wage slavery of factories and in the miserable homes and polluted air that were their lot.

A new embodied reverence for the energy of the earth requires not only a narrative sense for its origination in the earth's prehistory, but a sense of ecological connection between a furnace, or an air conditioner, and the fossils which fuel them. But the problem is that the machine age and the ethereal economy encourage precisely those habits of forgetfulness and waste which mean that modern consumers and corporations listlessly consume without heed for the past or future life of the creatures which composed, and will eventually have to deal with, the energy so used. A narrative and embodied ethic requires not only reverential rituals but a recovery of old virtues such as those of prudence and temperance, justice and patience, so that saving energy, consuming lower down the energy and food chain, substituting human labour for the machine – for example cycling or walking instead of driving – again appear hip and wise rather than old-fashioned and irrational (Northcott 1996: 314–18; Bouma-Prediger 1998). The virtues, as Aristotle first suggested, are those moral excellences which, when they become habitual, help to train the individual in relationships and roles which make for flourishing or what Aristotle called *eudaimonia*. The central and ordering virtue according to Christ and St Paul is love, love of God and love of neighbour. For Saint Augustine love is central and gives order to all other human desires, ends and virtues because when individuals love the being who has given them life, they turn their minds to the source of life. This source of life is the divine light which as St John put it in the prologue to his Gospel 'lightens everyone' and this image of divine light and energy provides the Christian account of the virtues with much of its metaphorical force. For Christians the ability to express the virtues arises not, as Aristotle has it from aristocratic upbringing and prowess on the battlefield, but from closeness to the ultimate source of goodness.

We rarely conceive of light from an electric bulb as the physical product of the light of the sun on the prehistory of the planet but this is precisely what it is. Embodying climate change will require us to behave more virtuously with respect to our individual and collective use of this divine light; it also requires us to reconnect the machine-enabled rituals of our lives with the ecology of our fragile and ancient earth home. Re-narrating energy as divine light, recovering the value of the traditional virtues, rehearsing the past lives which make our present ones possible, are practices which embody climate change and which may enable those who engage in them to give more cognisance to biogeochemical posterity.

References

Arhenius, S. (1896). On the influence of carbonic acid in the air on the temperature on the ground. *Philosophical Magazine, 41*, 237.

Bateson, G. (1973). *Steps to an ecology of mind: Collected essays in anthropology, psychiatry, evolution and epistemology*. St Albans: Paladin.

Bohm, D. (1980). *Wholeness and the implicate order*. London: Routledge and Kegan Paul.

Bouma-Prediger, S. (1998). Creation care and character: The nature and necessity of the ecological virtues. *Perspectives on Science and Christian Faith, 50*, 6–21.

Burke, E. (1790). *Reflections on the revolution in France and on the proceedings in certain societies in London relative to that event. In a letter intended to have been sent to a gentleman in Paris*. London: J. Dodsley.

Burkitt, I. (1999). *Bodies of thought: Embodiment, identity and modernity*. London: Sage.

Dupré, L. (1995). *Passage to modernity: An essay in the hermeneutics of nature and culture*. New Haven: Yale University Press.
Dupuy, J.-P. (2004). Intersubjectivity and embodiment. *Journal of Bioeconomics, 6*, 275–294.
Gibson, J. (1979). *The ecological approach to visual perception*. Boston: Houghton Mifflin.
Hauerwas, S., & Burrell, D. (1989). From system to story: An alternative pattern for rationality in ethics. In S. Hauerwas & L. G. Jones (Eds.), *Why narrative? Readings in narrative theology* (pp. 158–190). Grand Rapids: William B Eerdmans.
Landes, D. S. (1969). *Unbound Prometheus*. Cambridge: Cambridge University Press.
Lomborg, B. (2001). Global warming. In B. Lomborg (Ed.), *The skeptical environmentalist* (pp. 258–324). Cambridge: Cambridge University Press.
Mauss, M. (1973). Techniques of the body. *Economy and Society, 2*, 70–88.
Merchant, C. (1989). The death of nature: Women, ecology and the scientific revolution. San Francisco, CA, Harper and Row.
Merleau-Ponty, M. (1962). *The phenomenology of perception* (C. Smith, Trans.). London: Routledge and Kegan Paul.
Murphy, L. (1998). Institutions and the demands of justice. *Philosophy and Public Affairs, 27*, 251–291.
Nadler, S. (2011). "No necessary connection". The medieval roots of the occasionalist roots of Hume. In S. Nadler (Ed.), *Occasionalism: Causation among the Cartesians* (pp. 165–188). Oxford: Oxford University Press.
Northcott, M. (1996). *The environment and Christian ethics*. Cambridge: Cambridge University Press.
Northcott, M. (2004a). *An angel directs the storm: Apocalyptic religion and American empire*. London: I. B. Tauris.
Northcott, M. (2004b). The parable of the talents and the economy of the gift. *Theology, 107*, 241–249.
Northcott, M. (2007). *A moral climate: The ethics of global warming*. London: Darton Longman and Todd.
Northcott, M. (2010). The concealments of carbon markets and the publicity of love in a time of climate change. *International Journal of Public Theology, 4*, 294–313.
Phemister, P. (2011). Are mind-body relations natural and intelligible? Some early modern perspectives. In K. Allen & T. Stoneham (Eds.), *Causation and modern philosophy* (pp. 87–103). London: Routledge.
Sieferle, R. P. (2001). *The subterranean forest: Energy systems and the industrial revolution*. Isle of Harris: White Horse Press.
Spash, C. (2002). *Greenhouse economics: Value and ethics*. London: Routledge.
Strauss, L. (1995). *Liberalism ancient and modern*. Chicago: University of Chicago Press.
White, L. (1967). The historical roots of our ecological crisis. *Science, 155*, 1203–1207.
Zaner, R. M. (1971). *The problem of embodiment: Some contributions to a phenomenology of the body*. The Hague: Martinus Nijhoff.

Chapter 9
Make, Do, and Mend: Solving Placelessness Through Embodied Environmental Engagement

Isis Brook

Abstract How should we live in the world such that we have culturally enriching and worthwhile lives when the material and social fabric of our situation does nothing to nurture or sustain the kinds of relationships with each other and with nature that would seem to be a prerequisite for a healthy life? This chapter examines the claim that there are compensatory benefits – such as cosmopolitanism and increasing self reflection – that mitigate the psychological and social problems of living un-embedded lives in placeless environments. It then proposes the solution that simply by making things, actively engaging in things and, particularly, by mending things, we can rediscover the necessary environmental virtues to reintegrate ourselves into the material fabric of the world. Why this should work has to do with the transformatory power of active, purposive engagement with the material realm. Moreover, we can do this even in the midst of contemporary 'thinned out' spaces to make them into enriching places.

Keywords Environmental virtues • Placelessness • Transition towns • Sense of place

9.1 The Placelessness Thesis

With Edward Relph's seminal work, *Place and Placelessness* (1976), there has developed a strong normative proposition within place literature that says placelessness is a bad thing. What is placelessness and why is it a bad thing?

A common distinction within the literature is that between space and place, where space is empty of meaning and could be referred to by grid reference and place has unique characteristics that make it the place it is. Placelessness could be

I. Brook (✉)
Writtle College, Writtle, Chelmsford, Essex CM1 3RR, England, UK
e-mail: isis.brook@writtle.ac.uk

E. Brady and P. Phemister (eds.), *Human-Environment Relations: Transformative Values in Theory and Practice*, DOI 10.1007/978-94-007-2825-7_9,

thought of simply as standing in for the term space. However, it is more likely to be used when a place rich in local distinctiveness is destroyed by homogenising forces and turned into a space like many others. Typical examples would be when a street of individual shops is taken over by national or international companies and replaced with chain stores, or where a traditional market is covered over and turned into a shopping mall. In housing, the same shift can be seen where a national house builder rolls out the same estates with the same road layout and house types across the country; the only difference between the estate in Doncaster and the estate in Hemel Hempstead might be that the roads and cul-de-sacs of one are arbitrarily named after poets whereas the roads and cul-de-sacs in the other are arbitrarily named after British birds. Naming can be a powerful creator of meaning but this requires that the naming is, at least germinally, apposite.

So what's the problem? There are obviously efficiencies and positive benefits to these kinds of developments. What is claimed to be missing is the rich web of relationships that places and people will weave together. Out of these relationships a place emerges that is both nurtured by and nurtures the people who live there. When these rich relationships are truncated by some kind of wholesale development design from outside, the place becomes placeless and the people alienated from where they live (Hummon 1992; Lewicka 2011; Smith 2007; Twigger-Ross and Uzzell 1996; Buchecker 2009).

The displacedness of the contemporary urbanite means they are unable to care for their environment in the same way we do not care for a bland hotel room; someone else will clear up the mess and there is no reason to polish the furniture or nurture the corporate style planting scheme in the lobby. We either don't see that these things need doing or we don't think it is our job to do them. And if moved, as I admit I sometimes am, to surreptitiously dust the leaves of a plant with my own handkerchief to let it breathe such behaviour is deemed odd. There is a particular way we are expected to inhabit public space and caring for it with direct personal action is out of place. This public lack of care or detachment from our surroundings perhaps then plays out in the private realm of the home as well, where mobility works against investing love in objects, home or relationships (Pallasmaa 2008). Some forms of mobility, such as in refugee or exile situations, can induce strong attachments to the things brought with one; where all sense of home and personal identity can be invested in an everyday object. (Many of us will have felt a pale shadow of this in finding an item commonplace at home but exotic when travelling, for example the sense of home that a taste of Marmite™ can conjure. We are cosmopolitan travellers enjoying the strange but nevertheless hankering after the familiar.) With increasing mobility these attachments could become more and more individualistic, idiosyncratic and replaceable; weak stand-ins for the rich cultural meanings that develop through embedded belonging. Without that embeddedness we perhaps lose the ability to recognise, add to, and thus maintain and enrich places.

Therefore, a strong claim of the placelessness thesis is that, not only do we build dispiriting identikit towns but that these spaces build alienated selves. How and why this happens is put succinctly by Relph when he says:

> …we are creating, in Norberg-Schulz's (1969) terse phrase, a 'flatscape', lacking intentional depth and providing possibilities only for commonplace and mediocre experiences (Relph 1976: 79)

And presumably having only commonplace and mediocre experiences then has a direct impact on what we are capable of envisaging and of doing. Kathleen Raine in her autobiography gives a wonderful expression of this when she discusses the architecture of Cambridge and that of her hometown Ilford. She says:

> On the pavements of Ilford I had shrunk into myself in shame, physically present there against my will, elsewhere in spirit. Is not the chief pleasure we experience from architecture the change it imposes upon ourselves? The justification of the cost and labour of Gothic cathedrals and Renaissance palazzos is the greatness they confer upon, and demand of, those who come and go in them, inducing in us civilised modes of being and behaviour. The mean streets of the Ilfords of the world imposing meanness of thought, make impossible, or all but impossible, certain kinds of feeling, certain modes of consciousness; or drive these into bookish dreams. Conversely, in order to escape the silent demands of dignified and beautiful proportions, barbarians must desecrate and violate, smash the stained glass and deface the statues and paint defiant slogans on the walls that tell us all too clearly, in their beauty and harmony of proportions, that we might be better than we are (Raine 1975: 15)

9.2 Making Places and Made by Place

For its power and its positive agenda of resisting placelessness and promoting a fuller sense of human flourishing, the placelessness thesis requires the normative claim not only that we suffer in such places but also that places change who we are and what kinds of lives we can envisage and desire. We become, to follow Robert Sack's logic, 'thinned out people' and to be thus thinned out is a bad thing (Sack 1997). Alternatively, it might be claimed that this simply does not happen because, although we might not enjoy such places, they do not impact on who we are. Or, it might be claimed that while they do impact on who we are, this is in fact a good thing because we can then live quite happily in thinned out places.

More needs to be said about the nature of this relationship between people and place. Of primary importance is getting away from the idea that a relationship, in this context, is about two separate things interacting. Edward Casey makes this clear when he says:

> Any effort to assess the relationship between self and place should point not just to reciprocal influence ... but, more radically, to constitutive co-ingredience: each is essential to the being of the other. In effect there is no place without self; and no self without place (Casey 2001: 406)

This idea of co-ingredience can be used quite helpfully to undermine any idea that a place can be created merely by the addition of some meaning it has for someone. Jeff Malpas in his book *Place and Experience* takes to task those who link place to the idea of it being space plus its meaning or emotional connection to a subject. This would diminish the importance of place (as place and not just spatial location) as its existence would be entirely dependent on the perceiving subject. What Malpas seeks to do is to unpack why there is no experiencing subject without place rather than the other way around. He says: 'The crucial point about the connection between place and experience is not, however, that place is properly something only encountered 'in' experience, but rather that place *is integral to the very structure and possibility of experience.*' (1999: 32). This helps to usefully balance the commonplace notion that people make places with the more fundamental aspect of the relationship: that places make people. Thus when we ask questions such as, 'What kind of places do we want to maintain or bring about?, we are at the same time asking the question 'what kind of people do we want to be'?

9.3 The Nostalgia Critique

One of the major critiques of the placelessness thesis is that it promotes a backwards looking nostalgia for times past when people's horizons were thereby severely limited because they remained in the villages and towns where they had been born. This critique can be driven further by pointing to the stultifying atmosphere of social inclusion and the exclusion of people from elsewhere (Wheeler 1994; Leach 2002). Nostalgia can have these overtones. Moreover, it can be linked to the preservation of one story or image of the past and the eliding of others with the attendant power implications (Jackson and Penrose 1993), or simply the desire to maintain the visited other in a time capsule of one's own past (Massey 2007: 124). However, our ability to stand back from local custom and practice allows us to see these practices as local custom rather than how the world is. It is not that experience of wider horizons gives us a new access to truth it is just that different experiences allow us the possibility of self-reflection. For example, if I'm used to seeing the local harvest of apples gathered in round baskets that is how one gathers apples. If I visit another county and witness their local harvest of apples being gathered in wooden boxes I'm in a position to see that either will do. On returning home I might take a particular pleasure in seeing the round baskets again and enjoy the feeling that there is something right about gathering apples in baskets, but I can also wonder if boxes might be better. I might then realise that although boxes have certain advantages the fact that this county has a lot of willow and the other a lot of pine makes those differences understandable. In this instance local distinctiveness has come about not by arbitrary allegiance to local practice but by the availability of materials.

This is a very benign and easy example where the local practice is maintained for good reason, but what about other traditions? For example, what if our traveller comes from a place with a rigid caste system or female circumcision? In these examples the encounter with other social systems could bring about very helpful self reflection not just for the individual but for those suffering oppression when our traveller returns home. It is these kinds of examples of entrenched inequality that are presented as the down side of being embedded in place and local practice. Here the encounter with the other is seen not just as a different practice but ideas such as equality and self-determination are seen as universally better. As Yi-Fu Tuan points out practices as varied as: 'infanticide, child bride, scarification, bloody rights of animal or human sacrifice, foot binding, self-immolation of widows, demon possession, witch burning' add to the world's cultural diversity, but the loss of them 'can be borne by anyone touched by the spirit of enlightenment' (1996: 186). It is from this perspective that Tuan speaks of a high modernism and its ability to connect with both the local hearth and wider ideals such as rationality which he terms the 'cosmos'. Although he does suggest that in thinking reflectively we do separate ourselves from place and from our 'immediate group', what we gain is a connection to a wider world, to other places and other people. And 'a Cosmopolite is one who considers the gain greater than the loss' (Tuan 1996: 188).

9.4 The Compensation Thesis

This cosmopolite's viewpoint takes us to a further move in the debate: the compensation thesis. This is the claim, which has been in the literature on place from Relph onwards, that our alienated state does allow for and indeed makes possible a thoroughly modern and good feature of being human, that of freedom. Karsten Harries makes this point about our lack of rootedness:

> But once again: we must not lose sight of the gain in freedom this development has brought if the loss of place can be mourned, must it not also be welcomed as an essential part of the increasing emancipation of the individual from the rule of the accident of place? Place no longer need be destiny as it undoubtedly was for the Black Forest farmer. The more we want to emphasise freedom, the more likely we are to insist that emphasis on place should yield to a recognition of the value of open space (Harries 1998: 169)

Here the idea is that the thoroughly emplaced are living an authentic life but it is one prescribed by their situation. We need only imagine for a moment the divisions of class and gender to see how great a benefit some detachment from community, from place, from expected roles affords us as human beings living in the centuries that followed the enlightenment. In this sense the authenticity of the traditional for us has become inauthentic because what a human being is has changed; horizons have moved and to live a life prescribed by place is to no longer live a contemporary human life.

Casey discusses what he calls this 'compensatory logic of loss' (2001: 408). For him the human being is not undermined by the thinness of places because we gain so much from cosmopolitanism and self-reflection in the post-modern context. Moreover we can and do seek out thick places to further nurture our selves. The examples Casey gives of us of seeking out thickness, such as video/DVD not replacing cinema or Amazon not replacing bookstores (408), seem to me rather odd. He seems to be claiming that the continued existence of cinemas and bookstores is a marker of our continuing shared rich cultural existence. I can only imagine that Casey has neither been to a multiplex cinema recently nor asked a question that required some degree of literacy of a bookseller in one of the large stores like Borders. In many cities and towns in Britain, one can see only the latest Hollywood blockbusters and only those by going out of the town to a multiplex not easily accessible by public transport and watch such films along with an audience of about ten people on a smallish screen. This is not, I think, an enriching social experience.

Even for the ten cinema goers there, often teenagers, I suspect it feels more like something to do because there does not seem to be anything else to do. The idea that this replaces the pre- video/DVD cinema experience of going to one's local crowded cinema to see whatever film was showing that week, and would be a topic of conversation at work or other social events, is strange. This example, and one could do the same with recorded music vs the previous generation's norm of making music in the home or local venue, throws up a potential problem with the compensation thesis. The problem is that in the current stage of capitalist driven social structures a rich selection of meaningful social events is not actually available to people. Moreover, from the perspective of a 'thinned out' self it is hard to see where the motivation would come from to seek out such enriching experiences or the imagination to be the driving force initiating them? Capitalist structures are such that, contrary to the claims made, it is not the case that people get what they want, or what they deserve, they get what they are given – or rather, what they are sold.

Tuan's notion of high-modernism as set out in *Cosmos and Hearth* (1996) sounds convincing in terms of making the best of the freedom and self-reflection that displacedness affords us, but I wonder, if, like Casey's examples of seeking out thick experiences, it isn't ignoring the real nature of the thinness for most people. It is one thing to be highly educated and financially secure in a placeless environment and able to make choices about one's life, and quite another to be the recipient of a life of diminished opportunities in such an environment. The question that presents itself even more strongly in such a situation is, 'from whence comes the inspiration to engage in anything more enriching and meaningful?' And the post-modern response that whatever is being engaged with is obviously meaningful will not do because some engagements/activities just are, to use Mill's term, 'fecund' and some are not. In defining the characteristics of the higher pleasures Mill uses the term fecund in the sense they are richly productive of further pleasures. The reason, for Mill, that we should bother with the cultivation of such pleasures, or indeed such a life – rather than stick with the easy to obtain lower pleasures – is that they, once cultivated, are more pleasurable, as evidenced by our not wishing to trade them for lower pleasures (Mill 1962: 262). Meaningful activities do not just beget more pleasure/meaning but pleasures/meaning of a kind more worth having (Firth 2008).

Through some activities we learn to demand more of ourselves rather than of the world and it is these activities that we need if we are to redeem non-places, create healthy relationships and a sustainable wider environment.

9.5 A Solution via Matter

Help is at hand – quite literally. Placelessness discourse focuses on mobility and lack of belonging, but it could also focus on the way that mobility impacts on our day-to-day engagement with the world; how we do things here and now in the spaces, places and homes we inhabit. Alongside the historical shift to placelessness there has also been a move away from engagement with matter. This might seem a striking claim in a world we see accused of being ever more materialistic. But materialism as in wanting more gadgets or the latest technology and only being able to respond to the malfunctions of such gadgetry by discarding and relishing the opportunity to buy the latest model is not an engagement with matter and the material nature of the world, it is quite the opposite. And materialism as a position that denigrates any belief in the realm of the spiritual also casts the nature of matter in the wrong light. It helpfully rejects superstitions and dogma but can also reject any sense of wonder about the natural world and the construction of artefacts. Materialism in this form is, paradoxically, also a shift away from matter (Brook 2009). How can this observation help the placelessness issue? I think it can in a number of ways: primarily we can endorse or promote a 'new materialism' that really is about engaging with matter. Even if I live in a placeless environment if I can be encouraged to just engage with some aspect of the material realm I will be forging a new connection to where I am at the micro level of sitting here doing this and at the macro level of engaging with the material world with its attendant limits and balances.

In what follows, I look at four means of engagement encapsulated in the title: 'making things', 'doing things', 'mending things', and also 'making do'. I will regard these as environmental virtues and as the means to healthy habitation; healthy not just for the planet, but also for us through the self-transformative power of working with matter in an engaged way (Brook 2010). I focus on this latter aspect and leave as a reasonable assumption the view that environmentally good effects will follow. The relevance and applicability of a virtues approach to environmental ethics is discussed at length by Ronald Sandler (2007). He sets out how informed virtuous dispositions are action guiding and can deliver environmentally sound decisions.

9.5.1 *Making Things*

Although I recognise the problems set out by the nostalgia critique, the limits of space do not allow me to discuss that debate at length. As such, I will simply begin with the claim that in the developed world 'we don't make things like we used to'. Here I mean to point not just to the quality of things made but also to the way that making things has disappeared from our lives. If we do not work in a 'making things' profession then we are likely only to make things as a hobby; they are the products of leisure not the products of life. And just as many people used to make music or be able to cook from raw ingredients these are becoming lost skills now elevated to the realm of the arts or the quasi-arts of cottage industries that are so separated from our daily lives they become tourist attractions.

Adam Curtis in his documentary film 'Century of the Self' (2002) discusses an interesting example of initial resistance to convenience foods. He tells us that the psychologist Ernst Dichter advised the Betty Crocker Company that the reason why their cake mix was not selling was because it was too easy to use. Although it is perfectly possible to make a cake mix where no egg is required, as it can be contained in dry form in the mix, the use of such a cake mix was denying the cake maker the sense of having made something for their family. Dichter proposed that the mix should call for the addition of an egg. To him this meant that the housewife was symbolically offering up one of her own eggs to her husband and would thus return meaning to the cake-making enterprise. If we put to one side the Freudian overtones and a more exacting feminist analysis there does seem to be something very basic in Dichter's analysis of the situation. The cake made with the cake mix alone does seem to lack enough participation to be a satisfying experience and indeed cake mixes today still call for an egg to be added. However, why stop at the egg? If one has flour, sugar, butter, and eggs it is possible to make a cake, but not only a cake, one is then free to make any number of baked items from received recipes or those invented oneself. What is happening with the packaging of single items is the removal of both rudimentary skills and freedom. To return the requirement for an egg is merely the simulacrum of skill and engagement – it removes an instinctive disquiet about a distancing from the materials and actions of making something – without actually allowing freedom and skill development.

To look further at what is happening in the process of making something it might be best to turn to the accounts of craftspersons. Here it is possible to find very eloquent expressions that exemplify not just the production and refinement of specific skills but also a production and refinement of the self and more overarching virtues such as patience, an ability to move with the substance, the realisation of a gentle working with the grain producing more impact than forceful working against the grain. Also in developing craft skills is the meta-skill of self-reflection and evaluation without which the craftsperson can never move from apprentice to creative mastery (Kvale 2007). As Kvale points out the apprentice is traditionally working in an assessment rich environment where they see their work alongside the work of master craftspersons and more skilled peers and the response of customers – for example, does their bread sell. Here the apprentice baker learns about the customer, what goes down well in this community, as well as the materials of their trade. This kind of embedded skill acquisition is a long way from the labour of the production line. The personal creativity in honing a skill is evident in more traditional practices, although a resurgence of this can be seen in some more controlled or routinized employment, such as the practice of *la perruque* where a worker uses off cuts or wastage and the factory or office equipment and time to make something for themselves (de Certeau 1988: 25).

9.5.2 Doing Things

In my analysis of make do and mend it is in the making and the mending that the really archetypical actions are taking place but standing above that is the more general idea of action itself. This might seem to impose a false dichotomy between thinking and doing, and I would not want to suggest that in 'the new materialism' there is no place for thinking or that thinking is somehow to be avoided. To really engage with the material world requires an exacting and rigorous form of thinking because old abstractions and disengaged fantasies would no longer be the easy resting place for the mind.

What kind of doing would help placelessness? I am going to pick up one example of a burgeoning movement which seems to me to encapsulate many of my ideas here. This is the action being taken to produce transition towns (Hopkins 2008). Central to the transition initiative is the idea of a socially responsible way of responding to the impending crisis of peak oil (this is the tension between increasing demands from oil meeting dwindling supplies). Recognizing that individual self-sufficiency is never going to be an appropriate response, transition towns are those that have engaged in a highly public and shared discussion and debate about how their specific town is going to slowly adapt itself to a more environmentally sustainable future. Although focused on things like a carbon descent plan and ensuring environmental and community resilience this is just the kind of doing that could change what has become a placeless town into somewhere supportive of a rich interweaving of human and place relationship. It is interesting to note though that those towns that have quickly adopted this agenda, such as Totnes or Stroud, had never really made that initial transition into placelessness. This helps to underline the placelessness thesis that would expect soulless towns to lack the human movers and shakers who would have the necessary aspiration to begin the process.

9.5.3 Mending Things

It is a wonderful expression of care for objects and for the world that instead of throwing something away we mend it and give it, as the expression goes, a 'longer life' or 'a new lease of life'. Through our activity the object, which has fallen out of the realm of human use, is returned to that sphere and in that sense we do breathe life back into it. Previously a darned sock, reconditioned bike or glued together cup were a marker of poverty and perhaps they still are. However what we need to see now is the poverty of skill evident in the thrown away item that could easily be mended.

Contemporary manufacturing processes work against this with such things as designed obsolescence, repairs that are more expensive than buying a new item, modular components, sweatshop derived low prices, and upgrades that quickly out date previous equipment through lack of compatibility. These and many more of what an elderly relative used to call 'go-wrongary' built into modern products are what thwart us today in our effort to care.

However there are some areas of life where mending is still an option, and I want to take one of these as an example in order to outline what this kind of engagement with the world is like and what it can bring us. I am going to take one I am familiar with but other examples could be replacing the rotten ends of fence posts, or clearing away rust, treating, filling, rubbing down, and repainting the body work on a car wing. Many readers will have had the experience of a perfectly good and serviceable pair of jeans wearing through in just one area, if you are a gardener the knees and if an academic probably the seat. These can very easily be patched up by using a strip of some other fabric (perhaps even sacrificing one very old pair of jeans to provide fabric for this task for the next 20 years), a needle and thread. The task is not complex but does require a certain amount of dexterity and takes approximately 15 minutes. The wonderful thing about dexterity though is that it comes almost unbidden, you try to do something and in the trying you discover both how to do it and that you can do it better; it just requires a certain amount of tenacity.

But what about the fabric? In making the effort to engage with the initially recalcitrant denim and then finding ways to hold the patch and the jeans whilst manipulating the needle we develop an awareness of the nature of denim and the nature of wear and tear, we might extend the patch to cover not just the tear but to strengthen surrounding areas that are developing a soft plush of worn-ness. We might realise that jeans can be patched from the inside or from the outside depending on the type of damage and what is most comfortable to the wearer. The repair might incorporate a degree of darning (where the missing threads are replaced by stitching across the gap and weaving the replacement threads as in the warp and woof of the original fabric) as preliminary strengthening and to maintain the shape of the original prior to patching.

In hand skills there is always more that can be done with practice: more accuracy, smaller stitches, fewer visible repairs, faster execution, more creative solutions, fewer injuries, and so on. In the process of mending we put back the jeans and we develop our own skills, we join the jeans in a world of the repairable. C.K. Williams brings a number of themes together through the metaphor of mending in his poem 'Invisible Mending'(1999). Here Williams shows how the action and attendant attitude of mending does connect to some deep currents in ourselves. And with the repeated line "forgiveness and repair" (p. 70) he underlines the sense in which we practice forgiveness when we resist throwing away that which is no longer perfect but add to it our care to make it good again. Also we are better at mending something if we stand outside of ourselves and enquire 'what it needs' (Crawford 2010: 16) and, as Mathew Crawford points out, we need such breaks from our habitual self-absorption. In resisting the lure of the new and nurturing what we have, we practice a virtue that is self-transforming and exemplifies a care for the things in our orbit and a care for the wider environment.

As I said my focus is on the micro level and personal transformation but I will just mention that it takes 1,800 gallons (U.S.) of water to grow the cotton for 1 pair of jeans[1] and 73% of cotton is grown on irrigated land.[2] Hence the rather painful irony of the fashion statement of pre-stressed or ready patched new jeans.

[1] http://www.treehugger.com/files/2009/06/how-many-gallons-of-water.php. Accessed 20/11/10.

[2] http://www.panda.org/about_wwf/what_we_do/freshwater/problems/thirsty_crops/cotton/index.cfm. Accessed 20/11/10.

9.5.4 Making Do

I do not intend to say much about 'making do' except to use it to deflect any notion of impoverishment or suffering in endorsing 'make do and mend', just as the driving force behind the transition towns idea is not just a panic about dwindling resources or a mean-spirited notion of thrift. It is about saying we are not mourning the loss of our current lifestyles and making do with some diminished form of living. We are, rather, saying there is a lot that doesn't work about this twenty-first-century lifestyle for us as humans and the environmental crisis is just the trigger to get us communicating together about more enriching ways of living. Likewise, the positive aspect of making do is the invitation and endorsement of human creativity.

9.6 Conclusion

I would like to conclude with a specific example which I think exemplifies the make, do and mend idea. My daughter called in to visit after various trips away, and I noticed she was doing some crochet. I asked what she was making and she just lifted up her legs to show that she had only one slipper and said, 'I lost the other out of my bag somewhere and need to make another'. She had made the initial pair and was now able to replace a missing one just with her own ingenuity, craft skill and some spare wool. They might not be a perfect match but how much better to have, at least in some area of one's life, that degree of resourcefulness and mastery of materials such that the world comes close and we can accommodate ourselves to it.

By engaging with the material realm, not as automata, but as creative individuals we can do more with less and shape our lives and our homes and communities such that they exemplify enriching places of co-creation. In this way the seemingly intractable problem: of building or rebuilding nourishing places when the individuals there have been culturally impoverished, 'thinned out', by their current placeless environments, could be addressed. The answer to the question – where to begin – is right here, right now, wherever that happens to be. Therefore, my central thesis is that by endorsing make do and mend to encourage a form of immediate environmental engagement we can re-learn the skills and develop the virtues that will begin to redeem placelessness.

References

Brook, I. (2009). Dualism, monism, and the wonder of materiality as revealed through Goethean observation. *Philosophy Activism Nature, 6*, 33–39.

Brook, I. (2010). The virtues of gardening. In D. O'Brien (Ed.), *Gardening: Philosophy for everyone* (pp. 13–25). London: Wiley-Blackwell.

Buchecker, M. (2009). Withdrawal from the Local Public Place: Understanding the process of social alienation. *Landscape Research, 34*(3), 270–297.
Casey, E. (2001). Body, self, and landscape. In P. Adams, S. Hoelscher, & K. Till (Eds.), *Textures of place* (pp. 403–425). Minneapolis: University of Minnesota Press.
Crawford, M. (2010). *The case for working with your hands*. London: Penguin.
Curtis, A. (2002, May 1). There is a policeman inside all our heads: He must be destroyed. *Century of the Self.* BBC Four. Broadcast.
De Certeau, M. (1988). *The practice of everyday life* (S. Randall, Trans.). Berkeley: University of California Press.
Firth, D. (2008). Do meaningful relationships with nature contribute to a worthwhile life? *Environmental Values, 17*(2), 145–164.
Harries, K. (1998). *The ethical function of architecture*. Cambridge: The MIT Press.
Hopkins, R. (2008). *The transition handbook.* Totnes: Green Books.
Hummon, D. (1992). Community attachment: Local sentiment and sense of place. *Human Behaviour and Environment: Advances in Theory and Research, 12*, 253–278.
Jackson, P., & Penrose, J. (Eds.). (1993). *Constructions of place, race, and nation.* London: University College London Press.
Kvale, S. (2007). Contradictions of assessment for learning in institutions of higher education. In D. Boud & N. Falchikov (Eds.), *Rethinking assessment in higher education* (pp. 57–71). London: Routledge.
Leach, N. (2002). The dark side of the *Domus*. In A. Ballantyne (Ed.), *What is architecture*? (pp. 88–101). London: Routledge.
Lewicka, M. (2011). On the varieties of people's relationship with places: Hummon's typology revisited. *Environment and Behaviour, 43*(6).
Massey, D. (2007). *For space*. London: Sage.
Mill, J. S. (1962 [1861]). *Utilitarianism*. London: Wm. Collins.
Pallasmaa, J. (2008). Existential homelessness – placelessness and nostalgia in the age of mobility. In S. Bergmann & T. Sayer (Eds.), *The ethics of mobilities: Rethinking place, exclusion, freedom and environment* (pp. 143–156). Aldershot: Ashgate.
Relph, E. (1976). *Place and placelessness*. London: Pion Ltd.
Sack, R. (1997). *Homo geographicus*. Baltimore: The John Hopkins Press.
Sandler, R. (2007). *Character and environment: A virtue approach to environmental ethics.* New York: Columbia University Press.
Smith, M. (2007). Space, place and placelessness in the culturally regenerated city. In G. Richards (Ed.), *Cultural tourism: Global and local perspectives* (pp. 91–112). Binghamton: The Haworth Press.
Tuan, Y. (1996). *Cosmos and hearth*. Minneapolis: University of Minnesota Press.
Twigger-Ross, C. L., & Uzzell, D. L. (1996). Place identity processes. *Journal of Environmental Psychology, 16*, 205–220.
Wheeler, W. (1994). Nostalgia isn't nasty: The postmodernising of parliamentary democracy. In M. Pereiman (Ed.), *Altered states: Postmodernism, politics, culture* (pp. 94–109). London: Lawrence and Wishart.
Williams, C.K. (1999). Invisible mending. In *Repair* (pp. 69–70). Newcastle upon Tyne: Bloodaxe Books.

Chapter 10
Art and Living Things: The Ethical, Aesthetic Impulse

Reiko Goto Collins and Timothy M. Collins

Abstract In this paper we track the evolution of a specific 'ethical, aesthetic impulse' in environmental art practice, looking for indications of value exchange between the human and non-human. The ethical-aesthetic impulse is the move by artists towards nature as a context and subject for work that addresses ethical ideas and aesthetic opportunities in relationship to a changing environment. We outline a theoretical framework that is defined by environmental aesthetics and informed by ideas about relational and dialogic aesthetics. We come to the ethical issue through pragmatism and a specific approach that is grounded in phenomenology. This framework informs the analysis of three case studies of artwork with trees and forests by Alan Sonfist, Joseph Beuys, and Helen and Newton Harrison. Our analyses address the relationship between experience, imagination and memory as a pathway to empathic experience, which we argue is an important step in the evolution ethical values. In the final section, we consider how these artists have contributed to the freedom, health and wellbeing of trees through the range of spatial, social and political means that are available to them and describe how we ourselves integrate these ideas into practice in our project, 'Eden3: The Secret Life of Trees'.

Keywords Art • Aesthetics ecology • Empathy • Environment • Trees

Over the last 30 years a body of art work has emerged that explores creativity in relationship to land, environment and ecology. This can be described in terms of

R.G. Collins (✉)
Gray's School of Art, Robert Gordon University, 12 1/1 Vinicombe Street,
Glasgow G12 8BG, Scotland, UK
e-mail: r.goto-collins@rgu.ac.uk;

T.M. Collins
Glasgow School of Art, 12 1/1 Vinicombe Street, Glasgow G12 8BG, Scotland, UK
e-mail: timothymartincollins@gmail.com

E. Brady and P. Phemister (eds.), *Human-Environment Relations: Transformative Values in Theory and Practice*, DOI 10.1007/978-94-007-2825-7_10,

land or earth art, environmental art and ecological art practices. In short, the first move into the landscape was a strategic response to the limits of modernist-minimalist material, form and artmaking. Eschewing the gallery as the primary setting for art, artists began experiments with essential land forms, earth materials and related earth/sky phenomena. The result was a series of large scale sculptural actions and reactions to empty or open landscape and its related phenomena. Methods included digging, mounding, pouring and mark-making, often with industrial-scale tools and machinery. In a related but slightly different move, artists began to think about ecology and systems, complex inter-relationships of materials, natural phenomenon and living things in specific places. In this way of working, the traditional practices of marking and making gave way to experimental sculptural approaches: laboratory style environments and experiments that would reveal the detail, scale or complexity of living systems. While much of this system-based approach has an ecological/material focus which operates within an implicit social framework, some of the work has moved more towards an explicit social/ecological inquiry concerned with the ethics and values that shape the relationships between people, places and things. Below, we examine three examples of the latter approach with case studies about artists who are working directly with trees, with people and trees and public policy and trees. We conclude with a discussion of how we ourselves integrate such ideas into practice, explaining how this previous work informs the current development of our own artwork, 'Eden3: The Secret Life of Trees'.

Throughout this chapter, we track the evolution of a specific 'ethical, aesthetic impulse' in environmental art practice. This is the move by artists towards nature as a context and subject for work that addresses ethical ideas and aesthetic opportunities in relationship to a changing environment. Here, we consider the work as practitioners informed by philosophy, and we will conclude as practitioners informed by artwork. We come to this pursuit of the 'ethical, aesthetic impulse' through practice-led research and specific reflection upon previous experiences in Allegheny County, Pennsylvania. It was in Pennsylvania that we found ourselves trying to 'make aesthetic sense' of a changing post-industrial landscape.

Our point of view is defined by environmental aesthetics; particularly Arnold Berleant's idea of an engaged aesthetic and its intention to integrate subject, object and the field of experience into a singular aesthetic consideration. Our work is also informed by Allen Carlson's argument for a rational/empirical science-based approach to aesthetic perception (2000); and the integrative responses argued for by Brady (2003), Budd (2002), Eaton (2001), Saito (2008) among others. In our own practical application of this knowledge, Goto is closer to Brady and her integrative model, which focuses upon imagination and disinterest with limits on cognitive/scientific elements. Collins sits between Brady and Eaton: he sees scientific knowledge as an essential pathway to reveal invisible conditions and causalities that would seem to be essential to the challenges of contemporaneous imaginative response. Following Eaton (2001: 136–138) and Hepburn (2001: 61), we do not regard ethics and aesthetics as being normatively linked or ethical values as always present in works of art. However, we would suggest that ethics and aesthetics are relevant to critical analysis where stated as part of the artist's intent or where the critic finds ethics are implicit in the material, the subject matter, or the process and approach of the artist.

We have identified theoretical pivot points to inform our ethical- aesthetic inquiry. John Dewey provides a useful starting point, '....a person's ideas and treatment of his fellows are dependent upon his power to put himself imaginatively in their place' (Dewey 1934: 348). When we expand this 'fellowship' to other living things, we find issues that eventually led Goto to ideas of empathy and a deep reading of Edith Stein. Through theory, reflection on historic models and creative practice, Goto has been seeking intimate relationship and understanding of trees. Collins has focused upon German aesthetics and subjectivity, looking back to Schelling and Hegel for ideas about the integration of subject/object relationships, and the recognition of nature as the field in which human endeavour occurs. At the heart of the work is the search to see ourselves reflected in nature, rather than removed and opposed to nature. (This occurs with the understanding that all is nature, and what separates human and non-human nature are ideas reinforced by practice and tradition). We explore some of these ideas in our case study analyses below, as we search for indications of empathic relationships between humans and non-human living things, as well as a recognition of a value exchange that emerges from the integration of subject, object and the field of inter-relationship as the focus of critical aesthetic inquiry.

We chose 'empathy' as the framework for the analysis of our case studies. Following Edith Stein (Stein 1917: 14–22), we understand empathy as an act of perceiving in which we reach out to the other to grasp their state or condition. Empathic experience is about being immersed and open to something foreign rather than to something familiar. Empathy is not based on self-interest. It is a reaching beyond self without losing or forgetting oneself. In this way, empathy helps us to extend our own world image through interaction with different people and with other living things.

In empathic inter-relationship, trees do not have feelings, emotions and mobility like we do. They do have the ability to sense and to respond to light, temperature, water and the chemical components of air. Following the Santiago Theory, which Fritjof Capra attributes to Humberto Marturana (Capra 2002: 34–40, 42), *life* itself is a process of cognition – all living things have evolved the means to respond to perturbations in their environment. We have an ability to read the physical state of proximate living things. Through intimate daily relationship, we learn to recognize vigour, sluggishness or distress. It is possible to experience plants and trees empathically through careful observation. We share the environment with them. However, we do not share the physiological processes or spatial limitations of a tree; we do not perceive or respond to the environment in the same way. These are some of the things that makes trees foreign to us and at the same time very interesting and intriguing.

In the value exchange between people and trees, humans generate values from personal, local and expanded fields of self interest. In our work, we are looking for values that have been shaped by an expansion of human interest through human interaction with trees. We explore our ideas in the context of the specific artworks described in the case studies below and have chosen these works because each connects the generative potential of trees with emergent social, ethical and aesthetic interests.

10.1 Case Studies

In this section, we review three essential reference points through art created during the 1970s and 1980s in which artists work with trees or deal with trees as a part of an ecosystem. We begin with Alan Sonfist, one of the pioneers who opened up the idea of environmental ethics and of the artist's role in, and responsibility for, nature. We then consider Joseph Beuys and his approach to a culturally embedded symbolic meaning of trees. This is followed by a discussion of Helen Mayer Harrison and Newton Harrison who utilize a poetic and metaphoric method to reveal clear opportunities that are embedded in environmental complexity, political realities and international scale.

10.1.1 Case Study One

Time Landscape (1978)
Alan Sonfist

> The idea of digging up the past to bring it into the present is exactly what my art is about. I see myself as a visual archaeologist.
>
> (Sonfist 2004: 8)

Time Landscape (1978) is a permanent public artwork consisting of trees, shrubs and grasses at the corner of La Guardia Place and Houston Street bordering Soho, the historic arts district in Manhattan. The project site is 25 ft by 90 ft, (a former empty lot), which is separated from the street by an iron fence with no public access. In simplest terms, this is a sculptural idea (a forest) that has literally been planted and framed.

Sonfist's original proposal in 1965 included plans for over 50 forest projects in the region. After 13 years in development, Sonfist began working onsite in 1978. He researched local plants, geology, land formation and soil quality. He developed a plant scheme from species that were common to Manhattan Island before European settlers arrived in the seventeenth century. He planted beech trees that were grown from saplings transplanted from his favourite childhood park in the Bronx, a mixed hemlock and riparian hardwood forest that he talks about as a site of personal development and refuge. *Time Landscape* was a unique and visionary idea for its time, developed in the heyday of earth art and conceptual art.

Alan Sonfist is an artist who uses his memory, imagination and a natural science approach to make art. He embraces scientific knowledge and a cognitive/aesthetic truth that uses appropriate science and living reference ecologies to establish the forest. Sonfist's scientific knowledge extends and adds depth to his childhood experience and memories. The imagination that informs the work is nested in childhood experience and the embodiment of scientific knowledge over time. The reception of this artwork is more of a challenge as it is a complex piece that works on three levels. First, it is a sculptural idea that is also an evolving forest. Second, it questions traditional ideas about aesthetic objects, yet it retains and enforces a purely visual relationship: the iron fence provides both frame and boundary. Finally, the artwork encourages popular interest in the preservation of the historic fabric of Manhattan's buildings, to support the value and validity of preserving a historic forest in the city.

> Sonfist's success in persuading city planners and bureaucrats to approve the construction of *Time Landscape* is based on arguments that derive, not from conventional justification for public art, but from the discussion that surrounds issues of architectural preservation. Sonfist's stance has been that it is as important to preserve historical landscapes as to preserve buildings.
>
> (Eleanor Heartney in Spaid 2002: 7)

Time Landscape is a forest that benefits from the non-instrumental values it carries as a public artwork. Sonfist claimed derelict land that was then restored as forest and designated as a living object with intrinsic art/aesthetic value. Since it is not a public park, it is not subject to the instrumental values of parks and open space, and is therefore able to provide a unique and dedicated urban 'forest space'. This could only be achieved given the fact that it is indeed an artwork, not a park. Sonfist is quite eloquent on this point.

> Now, as we perceive our dependence on nature, the concept of community expands to include non-human elements; and civic monuments should honor and celebrate the life and acts of another part of the community: natural phenomena. (Sonfist 2008: 43)

Time Landscape is a symbol of forest and woodland which reminds us of the remnant conditions of nature in New York. It suggests an alternate aesthetic integrity. As a work unto itself, it has a core truth – an experiential truth that is confined to the realm of trees. This truth gains focus by its location, embedded at the heart of the city. At the same time, there is no indication that the work was intended as habitat creation, although the planting of diverse native plants communities that have then been protected from human disturbance, attracts and enhances urban wildlife, such as insects, birds and small mammals.

What would provide the imaginative authority to drive the development of such an artwork over an extended period of 13 years? An analysis of empathic interrelationship and value exchange may provide some answers. However, in order to understand these aspects of the work, we have to activate our own imaginations. It is clear that childhood experience with trees in a remnant landscape patch in the Bronx provided core experience for the adult artist. We think it is fair to claim there was value exchange in that childhood relationship to trees. The childhood experiences seem to have provided the creative authority and temerity to realize this work, just as they also underpin the imaginative act decades later. Goto argues that the 'aesthetic integrity' embodied in the work indicates a clear empathic relationship with the original Brooklyn site, as well as with the penultimate Manhattan site. In this latter case, she perceives an empathic relationship to a remnant ecosystem. She believes that the artist's world image has been extended by his original experience, then confirmed and clarified in the development of the work in Manhattan.

Sonfist's empathic relationship to nature is most obvious in the literature that attends his larger body of work. The source of that empathy is often discussed and developed through reference to childhood memories and the import of natural events within an urban setting (see Sonfist 1995: 158–159), while in a monograph on his own work, Sonfist makes explicit that his work is orientated by his personal history and his empathic/imaginative relationship to nature and its material (Sonfist 2004).

Time Landscape appeals to people who care about a niche for living things. At the same time, its presence provides an ethical, aesthetic counterpoint to normative urban speculation and development. It is a living forest sculpture that celebrates the uniqueness of an ecosystem.

10.1.2 Case Study Two

7,000 Oaks (1982–1986)
Joseph Beuys

> I think the tree is an element of regeneration which in itself is a concept of time.
>
> (Beuys in Kuoni 1990: 111)

7,000 Oaks was developed for Kassel, Germany by Joseph Beuys and presented at the 1982 international art exhibition Documenta 7. Beuys initiated his project by depositing 7,000 basalt columns in front of the Fridericianum,[1] the focal point of the exhibition and planting the first tree there for the opening. As the pile dissipated the forest expanded. On one level, this is a material/aesthetic artistic response to a city with a paucity of trees as a result of the Second World War bombing. On a metaphysical level, it is part of Beuys' larger project of creative social transformation. The exhibition opens with a transitional moment: the lone tree and the pile of basalt do not constitute an object or a concept that functions as a sculpture, but rather as a symbol of aesthetic change. The Documenta exhibition programme began in 1955, started by Arnold Bode, an artist and educator as a response to the fact that modern art had been prohibited under Nazi dictatorship.[2] Beuys's intention was to establish an international exhibition with contemporary artists that would play an important role in the regeneration of Kassel.

The stones and trees also have a deeper metaphorical function. Much of Beuys's work is influenced by the theosophy and teachings of Rudolph Steiner. His work was defined by ideas of freedom and the study of metaphysical truths that could be found in natural materials and their inter-relationships. He embraced materials that indicate warmth and energy and that suggest metamorphosis. The choice of the basalt columns is typical: it is a naturally occurring form that is curious in that it

[1] The Fridericianum was the first public museum in Europe, opening at the height of the eighteenth century enlightenment.

[2] Documenta [online]. [Accessed June 10, 2008] http://the-artists.org/tours/documenta1.cfm

occurs in common but not uniform shapes. Basalt columns are made by volcanic eruption and specific cooling conditions resulting in tension fractures which produce the unique columnar form.

As the tree grows, it is first overwhelmed by the density and presence of the stone, then it is briefly in material balance. After this, the tree overshadows the stone in a relational process that clearly marks the clock and calendar time of the viewer. Tree time, which Beuys estimated at 800 years for an Oak (Stachelhaus 1987: 149), plays a significant durational role in the work. Upon the death of the tree, the stone column (operating in geologic time) either marks what is missing, or encourages the city to plant another tree. The Oak was also chosen because it is a slow growing hardwood species and because it had long been a symbolic tree for the German people.

> We know that the oak in Germanic lands was the sacred tree, the tree of Thor, the god of thunder and lightning. Unless I'm mistaken, wasn't the tree also the symbol of justice?
>
> (Beuys in Kuoni 1990: 94)

The oak trees do not on first consideration appear complex. The oak itself has a powerful form, but the history, culture, symbol and myth underpinning the work are invisible to the eyes. Rather, they exist like stars in the daylight. Myth is constructed by the performative discourse that attends the work, realized through public pronouncements, interviews and actions. Kassel was once called Castellum Cattorum, a castle of the Chatti, who were a German woodland tribe known in ancient times as fierce hunters and gatherers who were also adherents of an animist religion. In a historical site not far from Kassel, Saint Boniface was reputed to have destroyed 'Thor's oak', a sacred tree of the Chatti, in 1737. (Talbot 1954: 45–46). Historical texts also describe the immense Hercynian forest, the natural landscape of these people extended from the Rhine to Romania – a forest that was a 9 days' journey in breadth and 60 in length. This history is an implicit but essential component of the artwork. The work can be understood as a material gesture of stone and trees with real impact for the City of Kassel, but it is also a neo-mythological return of the Hercynian trees and spirit to Kassel.

Beuys's work on the project is primarily embedded in an imaginative relationship with people, places and things. Can we consider the tree-stone body as the final form or is the generative growth and the potential for natural reproduction an important factor in the aesthetic sensibility of this work? The project was presented from the beginning as a narrative of recovery, as a transformative act that occurs in clock, calendar, tree and geologic time. If we were to view all 7,000 trees in 1 month, we still would not have 'seen' the work. Complicating things further, Beuys works from a rational historic understanding of super-sensible content. It is not cognitive/scientific as we traditionally understand these terms, but it does invoke cognitive/imaginative knowledge, and indeed this is essential to an understanding of the full scope of the work. An empathic relationship between human being and nature is constructed by a symbolic route. It is important to understand that to recognize empathy we must secure a symbolic relationship. The symbolic meaning of this work is embedded in a history of dwelling and livelihood amongst a great forest of trees. Beuys created and placed material symbols of a new urban mythology, 7,000 artifacts that facilitate empathic relationship between trees and everyday life.

Value exchange is embedded in the work's historic conditions and traditions. However, it is also embedded in Beuys's obsession with transition and transformation. The basalt calls the citizens to action, while the trees transform the city and establish the potential for a new forest culture. The stone/tree planting invite the citizens who pass by daily to witness the relational changes that occur over time. The stone symbolises both protection and witness and, in the final days of the tree, it serves as a clarion call for another generation to replant, if management schemes have prevented the setting of seed.

10.1.3 Case Study Three

The Serpentine Lattice (1993)
Helen Mayer Harrison and Newton Harrison

> We believe that our art works through metaphor – that all artwork is based in metaphor.
>
> Helen Mayer Harrison, in an interview with Reiko Goto (2008)[3]

In 1993, Helen Mayer Harrison and Newton Harrison (hereafter referred to as the Harrisons) exhibited *The Serpentine Lattice* at the Douglas F. Cooley Memorial Art Gallery of Reed College in Portland Oregon. The artwork is a multi-media installation that consists of large maps, slide projections, and three narrative stories about the tragic condition of the North American rainforest. The work explores what is generally understood about the health of the systems under scrutiny. It develops a metaphoric figure/ground narrative that reveals conditions, values and challenges, as well as a forward-looking vision. 'Serpentine' is the visual metaphor that describes the 2,000 mile long Western Coastal Mountain range which runs parallel to the shoreline. The 'lattice' describes over 3,800 watersheds, which drain from the north-south ridgeline to the Pacific Ocean in the west. The Harrisons have assembled eighty topographic maps from the United States Geological Survey into one large map. It is carefully hand coloured in emerald green with signs and symbols that tell a cultural, ecological, international, geological, topographical and sociological narrative of trees. The second component consists of over 300 slides organized to illustrate changes from pristine old growth forests, to select harvesting, then to clear cutting and roadways leading to destruction of habitat, slopes and watershed integrity. Aerial images show isolated patches and incredible fragmentation of the once continuous North American rain forest. All the images document the material condition of trees, the slopes, the remnant forest landscape. No pictures of wildlife, people, cities, towns, or any indication of human infrastructure were included.

[3]Helen Mayer Harrison and Newton Harrison, interviewed by Reiko Goto, 8 March 2008.

The Serpentine Lattice describes the loss of the North American temperate rainforest (the largest temperate rainforest in the world) and its dominant redwood groves. Images, poetry and metaphor are the media and method of this work. The project was underpinned by extensive rational analysis of impact and policy, based on discussions with ecologists, foresters, forest campaigners, scientists and woodsmen on the social, economic and ecological aspects of the forest. It is important to note that the project occurred in the midst of Bill Clinton's first term, a period during which renewed attention was directed towards the environment. The States and the Federal governments were both writing policy on the western forest. The forest industry was lobbying politicians. The presence of the Northern Spotted Owl prompted a closure (based on the federal Endangered Species Act) of National Forests to the wood industry. Environmental activism increased with 'Redwood Summer' characterized by talk of 'monkey wrenching'[4] and Julia 'Butterfly' Hill taking up residence in the branches of 'Luna' in defence of an old redwood that survives to this day.

A three part narrative was presented in the exhibition handout and catalogue. Excerpts were read by the artists during the lecture/performance at Reed College. The artists used conversational phrases such as 'Everybody Knows…' and 'Somebody Said…' to elucidate the range and depth of common knowledge. The title text synthesizes, clarifies and reshapes normative metaphors. In the conclusion, new ideas about forests, ecosystems, economic systems and public policy are introduced through a mix of careful analysis delivered through a reversal of the idea of the gross national product, presented as a 'gross national ecosystem', an indicator of an integrated social, political-ecological economy. At the heart of the project was a unique story:

> From southern Alaska To northern California
> North America's last great temperate rain forest is dying
> Everybody knows there's less than 10%
> Of the old growth left Between San Francisco and Vancouver Island
>
> (Harrison and Harrison 1993: 7)

By beginning with the words the 'rain forest is dying', the Harrisons evoke a compelling feeling of tragedy and concern for extraordinary trees. Our ethical sensibilities are engaged as we come face-to-face with the loss of a shared or common value. These narratives remind us of the conflict between the public and the private, the expert and the citizen. The work raises questions about the intrinsic and aesthetic values of the Pacific temperate rainforest that directly conflict with the practical utilitarian value of the trees as a natural resource. It challenges us to consider the relationship between knowledge of ecosystem loss and aesthetic experience of trees and forests and raises questions concerning the appropriate scale of experience: at what point do we actually 'see' this forest and its eco-social complications?

The Harrisons' work shows clearly that the forest had been compromised by an unsustainable rate of harvest and documents the severe economic impact this had on

[4]A term for environmental sabotage that was reportedly a methodology embraced by 'Earth First' activists. The term was first seen in Edward Abbey's book of the same name.

human life. Both the narratives and the slides provide us with avenues into an empathic relationship with a great forest, viewed as a living thing. Values are not exchanged *per se*, but rather are projected (as also in the work of Sonfist and Beuys) into the future. The normative metaphor in the American West identifies forest and land as fecund and regenerative resources. From the Harrisons' point of view, the forest canopy is the background or field upon which culture, industry and economy have developed with (minor or managed impact) in the foreground. By telling us the 'forest is dying', the normative metaphor is turned on its head. The images and text show us that clear-cut land, tree farms and road networks have become the dominant field. The forests that have driven both culture and economy are now (at 10% original cover) in the minority foreground. The work demonstrates the urgent need for restitution, for a return to the health of the forest, the great living thing.

By suggesting that the forest has been injured and is in need of restitution, are the Harrisons anthropomorphizing nature? We would say that this is not only appropriate, but it is an essential element of their strategy. Lakoff and Johnson suggest that 'perhaps the most obvious ontological metaphors are those where the physical object is further specified as being a person. This allows us to comprehend a wide variety of experiences with nonhuman entities in terms of human motivations, characteristics, and activities' (1989: 83). Characterization of the forest as a living 'thing' in need of restitution is an important metaphorical statement that supports the Harrisons' empathic vision.

The Harrisons have demonstrated a historic value exchange between people and trees, but one that, sadly, has lacked both wisdom and empathy. The value has flowed primarily in one direction, with short term benefit and long term impact. One reason to save the forest is to recover ecological values and biodiversity. The other reason to save the forest is to enable a healthy human relationship with this landscape, made manifest in a forest-based culture and economy.

The story of the dying forest concludes with the idea that 'Our eco-cultural entity can exist'(Harrison and Harrison 1995: 202–207). With this step, the metaphor or the generative ecosystem gains new scale and value. Nature and ecosystem are linked with human wellbeing and security. The Harrisons' imaginative proposal is that the Northwest rainforest is a 'gross national ecosystem'. This is an outrageous notion that nevertheless makes exquisite rational sense to all involved, for this is a metaphor that emerges from an empathic relationship which integrates the human condition, the ecology of place, and the wonder of living things.

10.2 Conclusion

In this final section, we make explicit the role of the trees in the historic artwork that has been discussed, before turning to speculate on how this informs experimental artwork today, especially our own. We understand that a tree, as a living thing, nourishes itself, through the organization and inter-relationship of its parts. 'Life' is defined by respiration, nutrition, excretion, growth and reproduction. Can we presume a baseline ethical, aesthetic position regarding living trees in this artwork? Do the artists discussed intend to contribute to the potential for trees to prosper, that is, to live, to grow, to reproduce and to age? Sonfist created an 'autonomous zone', a symbolic cultural place of natural potential. In Beuys's work, the strategic planting of 7,000 trees and stones resulted in a distributed spatial approach, a commitment to specific trees in place with intentional long-term consequence. The Harrisons made a sweeping aesthetic case for forest retribution. This was achieved through integrating trees within the natural freedoms of human beings, thereby extending moral obligation and public policy to trees on a national and international scale. This artistic/cultural approach to natural freedom, we conceive as the emergent ethical impulse. This impulse is a moral assignment that is normally only afforded to trees and forests that are exurban and of significant ecological import or to trees that occur on an endangered species list. These works constitute an artwork typology united by an experimental approach to moral inter-relationship with nature.

We have called this approach the 'ethical, aesthetic impulse'. However, what drives this impulse? We have sought out indications of empathic inter-relationship and value exchange between people and trees in these works. Certainly, there has been significant value exchange from the forests of Sonfist's childhood to the development of his adult work. This appears clearly in both the aesthetic and the intellectual intention of the work, but it is also corroborated materially. With reference to her own practice,[5] Goto claims that this type of work, including its obsessive integrity, could only emerge from empathic inter-relationship.

In the work of Beuys too, there is a clear value exchange with trees that is embedded as a thread that moves from forests, history and nature to artist and practice, and which then moves through the work to the daily experiences had by the audience. The audience's experience is also the best indication of the artist's empathic intent. Through their regular day to day access, the citizens of Kassel now participate in a temporal narrative of 7,000 trees played out against 7,000 stones in perpetuity. In the Harrisons' artwork, the dialogue is not with trees, but rather with the people who live and work with trees as part of their daily lives. The value exchange with trees is obvious in the dialogue with foresters, ecologists, activists and policy experts. There are clear indications of empathic inter-relationship in the materials chosen and developed for the exhibition. In effect, an unwieldy international scale topic of public policy is given emotional tension and access that facilitates empathic exchange with the viewer. In all three examples of world-class environmental art discussed here, we see an ethical baseline emerge. In each, it is underpinned by value exchange with trees, yet the empathic relationships that inform the work are harder to pin down than the empathic interface with trees that are developed in the actual experience of the works.

[5]Goto's work over the last 20 years has involved long-duration, inter-relational studies of trees, butterflies and their habitats, often with a focus on specific species. As well as shorter studies of a range of other creatures, she has created an extensive habitat on the roof of the Moscone Center in San Francisco.

Few alive today would disagree that humanity in the first decade of the twenty-first century faces the greatest set of environmental challenges that the civilised world has ever seen. Hundreds of artists around the globe have taken up the challenge of addressing issues of the environment. Some of these artists are conducting important experimental 'empathic' artwork with plants and trees. Those of whom we are aware include Agnes Denes (US), Yuji Dogane (Japan), David Dunn (US), Shelley Sacks (UK), Buster Simpson (US), as well as Natalie Jermijenko (US), who frames her work as objective science and technology, challenging the moral and ethical comfort zone of her audience in the process. Each of these artists is creating work that seems to be more deeply embedded in an experimental inter-relationship, interface and empathic exchange with plants and trees as part of the very development of the work itself. In the final dissemination and appreciation of each work, none of us has as yet had the impact of our progenitors to date.

As practising artists ourselves, we, the authors of this chapter, have been working on ecological issues since 1986. Our work has been deeply informed by the original contributions of the artists discussed above. Our research on nature and the post-industrial public realm, conducted over 10 years at Carnegie Mellon University in Pittsburgh Pennsylvania, resulted in 'Nine Mile Run' (1997–2000) and 3 Rivers 2nd Nature' (2000–2006).[6] Goto is currently reflecting upon the details of empathic interrelationship in this body of work.

During a short 3 month sabbatical in 2000, we spent a couple of days with scientists in North Carolina who were conducting long-term research into carbon dioxide and tree response. Standing at the top of a platform, just above the tree canopy the sun came up and the trees came alive, reacting to the first beams of light, then to a large diesel truck spewing carbon dioxide nearby. We were amazed as these seemingly inactive forest giants were transformed from sleepy to dynamic and reactive beings right before our eyes. This was the beginning of our thinking about working with trees. Eventually, in 2008, we began work on 'Eden3', a trans-disciplinary project involving artists and musicians, as well as philosophers, technologists and scientists, that seeks to both monitor and reveal the processes of respiration and photosynthesis in trees. We are developing a place-specific, sensual experience that links imagination and perception in pursuit of a means of imaginatively expressing the specific experiences of a living tree as it reacts and adjusts to the quantity and impact of carbon dioxide in cities. Goto references the human need to hear the breath of people, and of the other living things we care for, so as to assure ourselves of their wellbeing. This project, with its attendant technology, is a path to help us find a way to listen to the breath of trees. Only then will we be able to recognize that this is an essential reflection of our own breathing. While we may never see the reflection of our consciousness in trees and nature, the reflection of the essential spark of life is well within our grasp.

---

[6] Articles that examine the theory/practice intentions, methods and outcomes of these works have been by discussed at length by Lora Senechal Carney and Tim Collins in a special issue of the *Canadian Art Review* on the topics of 'Landscapes, Cultural Spaces, Ecology' (Collins 2010).

At the beginning of this paper, we began with a quote from Dewey: 'a person's ideas and treatment of his fellows are dependent upon his power to put himself imaginatively in their place' (Dewey 1934: 348). At this point, we are convinced that we must expand the circle of fellowship to include ALL LIVING THINGS . . ..

References

Beuys, J. (1990). Interview with Richard Demarco 1982. In C. Kuoni (Ed.), *Joseph Beuys in America* (pp. 109–116). New York: Four Walls Eight Windows.

Brady, E. (2003). *Aesthetics of the natural environment*. Edinburgh: Edinburgh University Press.

Budd, M. (2002). *The aesthetic appreciation of nature*. Oxford: Oxford University Press.

Capra, F. (2002). *The hidden connections*. New York: Doubleday.

Carlson, A. (2000). *Aesthetics and the environment*. London: Routledge.

Carney, L. S. (2010). Ecology and the ethics and aesthetics of collaboration. The case of nine mile run. In a special issue on *Landscape, Cultural Spaces, Ecology*, RACAR, Revue d'art Canadienne/Canadian Art Review, 35(3), 63–72. Arnprior, NO: Canada Universities Art Association of Canada.

Collins, T. (2010). *3 Rivers 2nd nature 2000–2005, Water, Land & Dialogue*. In a special issue on *Landscape, Cultural Spaces, Ecology* RACAR, Revue d'art Canadienne/Canadian Art Review, 35(3), 73–85. Arnprior, ON: Canada Universities Art Association of Canada

Dewey, J. (1934). *Art as experience*. New York: Berkeley Publishing Group.

Eaton, M. (2001). *Merit, aesthetic and ethical*. Oxford: Oxford University Press.

Harrison, H., & Harrison, N. (1993). *The serpentine lattice*. Portland, OR: The Douglas M. Cooley Memorial Gallery, Reed College.

Harrison, N., & Harrison, H. (1995). Newton Harrison and Helen Mayer Harrison. In B. Oakes (Ed.), *Sculpting with the environment* (pp. 198–207). New York: Van Nostrand Reinhold.

Hepburn, R. (2001). *The reach of the aesthetic*. Aldershot: Ashgate.

Lakoff, G., & Johnson, M. (1989). *Metaphors we live by*. Chicago/London: University of Chicago Press.

Saito, Y. (2008). *Everyday aesthetics*. Oxford: Oxford University Press.

Sonfist, A. (1995). Alan Sonfist. In B. Oakes (Ed.), *Sculpting with the environment* (pp. 158–159). New York: Van Nostrand Reinhold.

Sonfist, A. (2004). *Nature: The end of art*. Florence: Gli Ori.

Sonfist, A. (2008). Public monuments. In D. Montag (Ed.), *Artful ecologies* (pp. 43–44). Falmouth: RANE, University College Falmouth.

Spaid, S., & Lipton, A. (2002). *Ecovention: Current art to transform ecologies*. Cincinnati: The Cincinnati Art Center, Ecoartspace, and the Greenmuseum.org.

Stachelhaus, H. (1987). *Joseph Beuys*. London: Abbeville Press.

Stein, E. (1917). *On the problem of empathy*. (W. Stein, Trans.). Washington, DC: Institute of Carmelite Studies. (1989).

Talbot, C. H. (1954). *The Anglo-Saxon missionaries in Germany: Being the lives of S.S. Willibrord, Boniface, Sturm, Leoba and Lebuin, together with the Hodoeporicon of St. Willibald and a selection from the correspondence of St. Boniface*. London: Sheed and Ward.

Chapter 11
The Embodiment of Nature: Fishing, Emotion, and the Politics of Environmental Values

Andrea J. Nightingale

Abstract There has been recent interest in the emotional attachments to nature that appear to be bound up in people's willingness to protect nature. Within this work, the emphasis has been on the places and experiences people have in nature that help to form a sense of identification with their environment. This chapter seeks to build from these insights to explore how fishermen's embodied interactions with the spaces and places of fishing are integral to their understandings of 'nature' and 'fishing'. As fishermen move from their boats, to the pier, to the meeting room, embodied, emotional and importantly, political transformations in what it means to fish occur. I argue that these transformations are crucial to how fishers understand and seek to protect (or not) their fishing grounds and the political space within which they fish. These transformations are vital to understand in the Scottish context where there are increasing efforts to devolve fisheries management to fishers' associations, spaces where conflicts over how people care and seek to use the resource are profound and contentious. By exploring the embodied interactions that produce particular kinds of attachments to 'nature', new channels for fostering cooperation can emerge.

Keywords Nature • Embodiment • Emotional geographies • Political ecology • Scottish in-shore fisheries

The environmental movement has brought into the mainstream ideas about how to care, love and protect 'nature'. Notions of 'environmental values' are commonplace and almost exclusively refer to a set of values that privilege particular kinds of interactions and land management practices that are predicated on minimizing human impact or intervention with the 'natural' world. Yet, humans are deeply integrated

---

A.J. Nightingale (✉)
Institute of Geography, School of Geosciences, University of Edinburgh, Drummond St, Edinburgh EH8 9XP, Scotland, UK
e-mail: Andrea.Nightingale@ed.ac.uk

E. Brady and P. Phemister (eds.), *Human-Environment Relations: Transformative Values in Theory and Practice*, DOI 10.1007/978-94-007-2825-7_11,

within, and dependent upon non-human others in ways that belie attempts to leave nature untouched (Castree 2005; Cronon 1996b; Haraway 1997). In this chapter, I am interested in investigating how emotional attachments to 'nature' are linked to people's behaviours towards their environments. In a departure from other recent work in this area (Clayton et al. 2003; Milton 2002), I focus on people who work with natural resources, specifically fishing, for a living. How do people's embodied interactions with their resource and the larger networks of humans and non-humans involved in fishing create particular kinds of attachments and behaviours towards their environments?

This work arose from preliminary research on Scottish in-shore fisheries management and my long-term interest in theories of nature and the commons. My other work on common property seeks to understand the importance of social relations and antecedent institutions and practices on the emergence of management regimes for the commons (Nightingale 2011a, b). My interest here is in expanding this work to begin to take account of affective relations that I argue are integrally bound together with these other elements of socionatures. A key way in which I explore these relationships is through embodied practices and the relationship between bodies, environment and emotions. Central to my theoretical purpose is to illustrate how the boundaries between environments and the subjective experiences of individuals and groups are not stable and this instability highlights the relational production of socionatures. Attention to how such boundaries are constructed and resisted helps open up conceptual space to think about the importance of emotion in embodied, social and ever dynamic relationships with the non-human world.

To set the stage, I need to explain more clearly what I mean by 'natural' or 'ecological environments'. I understand environment to include anything outside of the self and is therefore inherently a fluctuating concept. Work on agoraphobia has shown that what is considered outside of the self is not necessarily straightforward or stable (Davidson 2000, 2003). This suggests that the 'self' is not essentially contained within the body or the psyche, but rather is something that emerges relationally in specific contexts. Even in a 'purely' material sense, the body does not neatly separate us from the 'outside' as we host bacteria and ingest food through the alimentary canal. In addition, I take seriously insights from nature-society geographies that explore the problematic construction of 'nature' as something opposed to 'society', and rather treat nature as a social concept that is culturally and historically specific (Braun and Castree 1998; Castree and Braun 2001; Cronon 1996a; Hinchliffe 2007; Whatmore 2002). Nevertheless, I will here focus my inquiry around natural environments—that which is taken-for-granted as 'nature' by most people.

My use of the term 'subjectivity' requires some clarification. Subjectivity is often conflated with identity, but here I distinguish between a conception of the subject and the identities people embrace and enact. The subject is constituted by power, both power over and the power to act and often refers to the discursive ways in which people become subjects of states or other types of authority (Allen 2002; Butler 1997; Foucault 1991; Henriques et al. 1984; Malone 2000). It is also used to understand the operation of power in society more generally (Butler 1997; Gibson 2001; Mahoney and Yngvesson 1992; Probyn 2003; Scott 1991). Feminist theorists often refer to the ways in which people are 'hailed by' or subjected by subjectivities such as gender and race which while they may resist, they find very difficult to escape (Bondi and Davidson 2003; Gibson 2001; Longhurst 2003; Probyn 2003). In this paper I use this kind of understanding of social relations to explore how fishers are enrolled into particular kinds of subjectivities that help to bound the values and (embodied) interactions they exhibit both with the sea and the political-economy of fishing.

11.1 Scottish In-shore Fisheries

To ground this discussion, a brief introduction to my work with the Scottish in-shore fishery is in order. This chapter is in part the outcome of preliminary ethnographic work on the Scottish in-shore *Nephrops norvegicus* fishery. *Nephrops* are also known as Norwegian lobster or prawns and are the main species marketed as scampi or langoustines. They are fished both by creel and by trawl net, although the creel fishery produces a higher value, live product. *Nephrops* are crucial to the Scottish west coast fishery as they account for the vast majority of landings (Scottish Executive 2005). The west coast is a mixed fishery with creelers and trawlers sharing the same fishing grounds in most areas, with significant salmon and mussel aquaculture developments in the lochs that overlap spatially with the wild fishery.

The European Union Common Fisheries policy sets prawn quotas and the UK government distributes them to set limits on the Total Allowable Catch in (for example) the prawn fishery for the west coast of Scotland. There are very limited opportunities for fishers to self-regulate (Scottish Executive 2005). One community on the west coast has banned all mobile gear (trawling) from their fishing grounds and they operate a formal, although not legally binding scheme to limit the number of creels fished per day per boat. They are a very interesting case study as the scheme was initiated by the fishermen in a political context where such schemes are not encouraged or supported and in some cases are illegal (Crean 2000).

In a number of places, skipper-owned export companies export prawns to the live prawn market in southern Europe, although many export companies are owned by non-Scottish interests. Some prawns are consumed in Scotland but the vast majority are exported directly (and, in the case of live prawns, on the same day) to southern Europe.

The in-shore fishery is facing significantly different pressures than the highly publicized white fish fishery. The decommissioning of boats and limited days at sea for the white fish fleet have put more pressure on the in-shore fishery as many boats are moving to the Scottish west coast seasonally in order to get more days at sea. Nevertheless, the fishery continues to be dominated by skipper-owned boats that are operated out of small ports on a daily basis. It is rare for skippers to stay out at sea for extended periods of time, although some trawlers will go for up to a week. The fishery is also dominated by men. There are some women who fish, but the vast majority of boats are skippered and crewed by men. This context provides fertile ground for exploring the boundaries between self and other, values and environment and the consequences of that for the relational emergence of nature and society.

11.2 Embodiment, Subjectivity, Emotion and Environment

The relationship between embodiment, subjectivity, emotion and environment is one which requires an interdisciplinary exploration and the forging of new ground. There has been some work done in anthropology and sociology on emotion, identity and environment which forms an important starting point for my argument (Clayton et al. 2003; Milton 2002). Kay Milton (2002) has examined how emotions and identification are part of forming values and attachments to environments and play into how we perceive and use the environment. She has argued that if people can identify with aspects of their ecological environment as being 'like' themselves in one way or another, they are more likely to treat that environment as they might themselves or another person. Identification leads to a set of values that involve care and respect. I move away from an explicit focus on identification and rather think about the process of subject formation and how that process is embedded within and indeed inextricable from both relations of power and the socio-natural environments within which subjects emerge. This kind of conceptualisation allows individuals to inhabit more than one kind of relationship with 'others', including nature, and thus to better capture why people's behaviours and values can seem to contradict the very values they are working to preserve—for example, a deep attachment to fishing yet engaging in fishing practices that lead to over fishing.

I turn now to the west coast in-shore 'fisherman' and examine three fishing 'moments' in an attempt to animate my more abstract points. Through this discussion I first illustrate the relationship between embodiment, emotion and subjectivity, and then show how emotion is bound up in people's particular engagements with fishing policy, practice and regulation.

11.2.1 Fishing the Sea

Fishing produces particular kinds of subjects and bodies. To be a 'fisherman' is first and foremost a masculine subject in Scotland and one which requires going to sea. It is not possible to be a fisherman and not fish. Yet the boundary between fishing and not fishing is not as clear cut as it may at first appear. I spoke at length with a man who went to great pains to justify to me that he is a fisherman even though he no longer goes to sea due to problems with his back. Instead, he is now working on fisheries policy issues but does this as a 'fisherman'. This man had a strong emotional attachment to being a 'fisherman', something which transcended the actual act of fishing and being on the sea, although its origins were in the 20 years he spent fishing.

The only way he can claim an identity as a 'fisherman' is to highlight the years he spent fishing. Yet in many respects his subjectivity has changed substantially now that he no longer works on boats. He spends his time working in policy arenas, dressed in suits and filling in paperwork. These embodied tasks place him in relation to 'fishing' very differently, and I believe it is because of this that he is so insistent that his time on the sea still qualifies him as a 'real' fisherman.

In fact, lest I doubt his intentions in working for protective policies, he told me, 'I am not an environmentalist at all. I couldn't be as a fisherman.' He said this in the context of a long discussion about his attempts to institute a regulating order for the west coast. The regulating order would have allowed fishers to manage the west coast more locally rather than from Edinburgh, and probably would have led to more restrictions on fishing, not less. Yet he wanted me to be absolutely clear that in his mind there was a distinct divide between those who seek to protect the sea (environmentalists) and those who manage/regulate the fishery (fishermen).

This attachment to the subject 'fisherman' created a set of values that led him to invest his time and energy into creating in-shore fisheries restrictions. The regulating order was not successful as the Scottish Government has implemented Inshore Fisheries Groups (IFGs) instead which are another mechanism to devolve management (Scottish Executive 2005). At that political moment, therefore, the regulating order was superseded by the IFGs. This decision came after my interviewee had spent the better part of 10 years in cooperation with others trying to develop the regulating order. Despite this set-back, he believed the regulating order was the best way to protect the fishery and to ensure that local fishers could negotiate with buyers for the best prices.

To be a fisherman then, is something that requires one to express a set of values and engage with a different network of relationships that are distinct from the ones 'environmentalists' are engaged in. And yet, in some respects, the goals of both networks are the same: to keep fish stocks at sustainable levels. In the eyes of many environmentalists, this requires fishers to stop fishing all together. Fishermen, in turn, take pains to insist that they are not environmentalists even if they are working equally hard to keep their catch sustainable. They value a working environment, use of the sea and reject the environmentalist stance that generally leads to the conclusion that all fishing needs to be halted.

The problem with this kind of polarization is that it fails to take account of the practices in everyone's daily lives that feed into fishing as a problem. Few environmentalists swear off fish all together, and even if they do, they simply spread their dietary needs to other kinds of resources—often vegetables produced on contract farms in the third world (Little and Watts 1994). Fishers find it difficult to relate to environmental movements because they are the ones cast as the problem, rather than the complex set of global geo-political economic relations that create particular kinds of fishing technologies and markets. In this case, the west coast fishery is closely tied to shellfish markets in southern Europe. At the moment, the demand for a live product has helped create the market for creel-caught fish which by all assessments is far more sustainable biologically than trawling (Atkinson, personal communication, 2007; Catchpole et al. 2005; Ridgway et al. 2006; Tuck et al. 2000).[1] Fishing practices, markets, fishermen and the biology of the fish, therefore, all emerge relationally and cannot be neatly separated from each other.

Here, the links between emotional attachments and political economic power relations come clearly into view. This particular man is deeply attached to being a 'fisherman', an attachment that is clearly emotional (as well as political) even though he no longer goes to sea. That emotional attachment then places him within particular kinds of subjectivities and relationships that have important effects on how fish and fishing policies circulate. He was operating at a Highlands-wide scale trying to bring people together across the west coast to agree to the regulating order, at the same time negotiating in Edinburgh with the Scottish Executive. If he had been an 'environmentalist' the effect of his lobbying for a regulating order would have been very different. In the eyes of the fishers, it would have had no legitimacy, and how he would have been received centrally is also dubious. More than likely no one would have believed that he represented fishers. I want to make it clear, however, that he *could* have chosen to interact with fishing policy as an environmentalist and it is on that basis that I suggest he has a strong emotional attachment to being a 'fisherman' and set of values that places him within the 'sustainable use' camp rather than 'protectionist' camp. I have met other fishermen who draw a less rigid line between the two positions. That, and the ways in which he sought to justify to me that he was a fisherman suggests that he was very proud of having fished and if he could, would still be fishing now.

One question that arises from this example is this: how do we make sense of this link between emotion, values and action? I suggest there is a particular sense of self and sets of relationships that allow one to fish, and these are then relevant for how the same people come to see themselves and their actions in relation to environments. In other words, it is the embodied act of being on the sea and fishing that is important in producing strong emotional ties that then translate into political relationships. What is under-theorised here is the link between embodied action and emotion. For that we need to turn to feminist theory.

Within feminist theory, a lot of attention has been paid to the production of the subject, or the ways in which people come to be subjected by particular kinds of discourses (Bondi and Davidson 2003; Butler 1997; Longhurst 2003; Mahoney and Yngvesson 1992; Probyn 2003; Wilson 2001). Judith Butler has taken this a step further, and explored the contributions of psychoanalysis to understanding the attachment of the psyche to being subjected (Butler 1997). Such work has much to offer to an analysis of environmental issues since whether people comply with institutionalised rules for conservation, I suggest, is integrally bound up in daily contestations over identity, self and other. The attachment to being in relation (or in society) makes the subject 'turn back on itself' and seek subjection. This is a multifaceted

[1] Professor Jim Atkinson and colleagues have done recent research in the inner Minch of Scotland (west coast) that shows morality rates from creel discards is substantially lower than discards from trawl nets. In addition, they have done experiments on 'ghost fishing' of creels and found that creels abandoned or lost on the sea bed do not continue to fish but rather become dens for crabs.

process whereby the subject (re)produces its own subjection even when it actively resists that subjection. In other words, when one promotes 'responsible fishermen' schemes one is simultaneously reinforcing the notion that fishermen are not responsible. It is only by accepting that one is subjected to particular discourses that one can resist that domination (Mahoney and Yngvesson 1992; Scott 1991). Butler (1997) therefore describes 'the subject' as 'the effect of power in recoil' (p. 6) and argues that 'in the act of opposing subordination, the subject reiterates its subjection' (p. 11). Thus subjects are always already embedded within relations of power, relations which cannot be overlooked when exploring how people come to care about and seek to protect their environments.

Building from Butler's more abstract work, within geography attention has been paid to the embodied, discursive and social processes that produce subjectivity (Gibson-Graham 2002; Gibson 2001; Longhurst 2001; Probyn 2003; Rose 1993). Longhurst (2003) outlines three main theoretical developments around the subject within Geography. First, subjectivities are always placed, in other words subject positions do not exist in the abstract, devoid of context, but rather are performed within specific places and spaces. Examples from the literature include the gym, the typing pool and the aeroplane aisle (Bondi and Davidson 2003); and here I would add, the fishing boat. Second, subjectivities are embodied and as such are material. Work on gender, race and ethnicity is particularly salient as it shows the ways in which people's experience of being a physical person is bound up in how they are subjected (Longhurst 2003). As I elaborate in the next section, the materiality of fishing and fisheries policy contexts are constitutive of fishing subjects. Third, the politics of subjectivity have been brought into view with particular attention to resistance and the ways in which resistance is always contradictory and often paradoxical (Longhurst 2003; Mahoney and Yngvesson 1992; Nightingale 2006). In the case of fishermen, they can only work towards various kinds of 'sustainable' fisheries arrangements by first accepting that they exploit the sea. They are 'not environmentalists', meaning that of course they use the sea. Their work towards different kinds of fisheries agreements would have a very different meaning if they were already accepted as wise stewards of the sea.

The kinds of subjectivities people are 'hailed by' (Probyn 2003), therefore, are important for their values, political networks and the implications of their actions in those networks (Gibson-Graham 2002; Gibson 2001). But political action is not really sufficient to explain why people become passionately attached to different kinds of environmental action. Rather, we need to understand how embodied experiences produce emotional attachments to particular environments and particular subjectivities, a cocktail which develops into a strong set of values that then place individuals in relation to the fishery in particular ways. What emerges is an understanding of subjectivity and space that insists that subjectivities and space are interactive and mutually constituted and have implications for values and environments.

Theorisations on subjectivity and space thus begin to open up questions about how environments and selves are co-produced and the network of relations that sustain subjectivities, yet ecological environments have been notably absent from much of this work. Early theorising in feminist geography insisted on the role that social relations play in the production of space and place (Bondi 1993; Hanson and Pratt 1995; Massey 1994; McDowell 1983), but this is a different argument from theorising how space, place and nature produce subjectivities. In addition, this work has not explored issues of boundaries and how the boundaries between self and other are implicated in the processes of subject formation. More recent work on subjectivity and space brings to the fore questions of boundaries and how subjectivities are bounded by and in turn bound space and place (Bondi and Davidson 2003; Longhurst 2003). Work on agoraphobia in particular has been useful in thinking through boundaries as many people suffering from agoraphobia describe an inability to draw a clear boundary between themselves and their environments (Davidson 2003). I suggest that the kinds of boundaries that fluctuate in fishers' lives between themselves, the sea, the community and fisheries managers are crucial to the kinds of values and politics that emerge.

11.2.2 Embodiment, Emotion and the Fishing Subject

The second fishing 'moment' I use to expand upon these ideas is the meeting room. In-shore fishermen spend most of their working lives on small boats, on the sea, hauling in either trawl nets or creels. The work is dangerous, physically demanding and generally requires long hours at sea. There is a very macho culture around fishing, with some fishers competing with each other over who can do longer days or haul more creels faster. However, the point that needs to be emphasized here is the embodied nature of the fishers' experience of the sea. While they do not go under water (unless diving for scallops), they nevertheless have a particular understanding of the sea that derives from their experience of the waves, the water both on and below the boat, the composition of the catch and observations from the boat. Much of this understanding is not even conscious, but rather something they learn over time.

Work on the boats then is integral to the production of the subject 'fisher'. It is the embodied act of working on wet, smelly, cold and dangerous boats that is important in creating a boundary between the subject 'fisherman', 'community' and the 'sea'. Many fishers spoke of their love of the sea and the need to have a career at sea. One fisher who is fed up with fishing told me he hates fishing, 'but I can't imagine a career that didn't involve going to sea.' For him it is not even the subject 'fisherman' that he is so attached to, but rather the sea itself. He was eager to be involved in my research as he saw it as one way to make 'fishing' more interesting. While I have yet to explore this in any detail with fishers, here I would suggest that for many, they see themselves as 'part' of the sea in some sense. The sea is a key defining feature of their lives and when a fisherman retires, they refer to his change of location as 'coming ashore', even though these same fishermen are generally on shore every night. Certainly, they make a distinction between people who work on the sea and those who do not. This same fisherman's wife asked me when I arrived at their house if I 'was another one of those crazy people who's in love with the sea.' In many respects, this man's sense of self is shaped by the sea.

The boundary between fishers, the sea and community, however, shifts rather radically in the meeting room. Sitting in unfamiliar suits or in jeans (with policy makers in suits) produces a very different kind of 'fisher'. Here, fishermen are often blamed for over-exploitation of the sea or told what kinds of regulations they must work within. Their role in the community shifts from one of provider and economic anchor of small coastal villages to one where they have to justify their occupation in a variety of ways. At one meeting, fishers who had lobbied for a long time to impose restrictions on the boats in their area were told that they would probably have to wait until the Inshore Fisheries Groups (IFGs) were finalised. They would then need to begin negotiations with the IFG over their proposals. The meeting room, therefore, puts fishers in a very different relationship to their resource. They have to justify their actions, are told *they* are over-exploiting the resource, or in this case, that their own ideas about how to manage their local fishery need to be subsumed to larger, national level processes. It also produces very different emotions about the process of fishing as they find themselves needing to fight not only conditions on the sea, but also the political climate. Importantly, fishers sense of self shifts in these contexts as relationships with the sea, the community and fishing emerge emotionally and materially to produce different subjectivities and politics.

To take stock, so far I have tried to argue that for some fishers emotional attachments to the sea are important for their motivation to be involved in various regulatory or management initiatives. This shows the ways in which emotions are clearly important to and embedded within political economic processes and social relations that extend far beyond the individual. Thus, following Liz Bondi (2005), I argue that emotions are relational and not something that are held within individuals. Rather the kinds of emotions people feel and display are deeply social, products of relational, embodied interactions with both human and non-human others and form a foundation for their values towards environments. In some sense emotions flow within interactions, rather than being held by individuals (Bondi 2005). Emotions are not separate from values or power relations, but rather emerge within power laden interactions and subjectivities. Thus, fishermen's attachments and values towards the sea are bound up in the kinds of interactions they have with the sea and the regulatory environment of modern day fishing as well as other aspects of their personal and social experiences (Brady 2006).

Returning to the sea, the boundaries between what is self and other that are created through emotional processes are in turn embedded within the kinds of actions that people take. Boundaries are important in differentiating not only 'fishermen' from 'environmentalists' but also in separating the 'brave, providing fisherman' from the 'exploiter of the sea'. These kinds of subjects are highly political and have real consequences for how fisheries management plays out, particularly as the same individuals often move between these positions. It is difficult for fishers to want to abide by regulations when they feel that they are simultaneously blamed for the need for such regulations and yet not allowed to create their own agreements. In other contexts, their attachments to the sea and the community provide a strong counter-motivating force for self-regulation.

Note that I am not equating subjectivities with identities. People move between subject positions, often in fluid, contradictory and unpredictable ways which is different from identities which tend to be more stable discursive constructions (Nightingale 2006). By emphasising subjectivity, issues around people's sense of self and how that emerges relationally from embodied experiences is clearer. This conceptual move then opens up space to think about the relational emergence of emotions, subjectivities and environments.

11.2.3 Attachments, Values, and Environment

I now turn to my third 'moment' so as to bring together some of these ideas and illustrate them a bit more fully. One fisherman I spoke to resonated strongly with the idea that attachments to the sea were important for fisheries management. In response to my statement to that effect he said,

> Interesting you say that. Right now there's a lot of pressure on the stock in this area…The bay out here, we used to fish it for part of the year then move off and go elsewhere. Now creels are constantly there. I haven't fished there for five years…We're light years behind the guys up north [who have a fishery closed to mobile gear]. They have their closure. Two boats here sell through [their export company] and we follow their practices: we don't take berried prawns, no small prawns. We're doing it because it feels good. You feel you're doing something.

For him, 'feeling good' was linked to managing the fishery sustainably. Not only is he attached to the sea, he is also attached to a sense of treating the sea well. He felt frustrated that while most of the fishers in his area shared his view, there wasn't a better legal structure to facilitate their goals. His discussion of attachment to feeling good segued into a discussion about trawling. He outlined the problems with trawling, particularly big trawlers that cover much of the west coast and said, 'Large trawlers are a business as opposed to a way of life' He contrasted that with creeling which he said was far more than a business, 'Days you're out there and you're barely making a living but you're at sea…It's a way of being.'

This fisherman draws a clear distinction between the smaller boats (in-shore) and larger trawlers who work both in and off-shore to make a point about how they interact with the fishery differently. Not only is the trawl catch more indiscriminate, but he also suggests that their emotional attachments to the sea are dissimilar. It isn't a 'way of life' for them, it's 'a business' implying that they are only interested in making money. They do not have the same commitments to try to manage the fishery sustainably because their emotional involvement is different. He also pointed out that large trawlers do not spend money in the village and have no commitment to the community. For him, there was a fuzzy boundary between feeling good, fishing as a way of life, and giving to the community.

His discussion of trawlers and their large by-catch (unusable, dead fish) led him into a long story about how he felt when he pulled up a minke whale dead that had become entangled in his creel line. It happened over 10 years earlier but he was still deeply disturbed and vividly pictured the whale as it emerged at the side of the boat. He concluded his story by saying, 'It's important to me to feel good about what I'm doing everyday. That's why I'm working for a closure in this fishery.' Clearly, there is a strong link between his emotional attachments to the sea, his values and his political work in fisheries management. In this context creeling seems to produce different attachments and values towards the sea than working on a large trawler.

I suggest that this is due in part to the embodied nature of the work. On a creel boat you are much closer to the water, you are working with live fish (most fish are dead in a trawl net), and importantly, creel boats tend to operate within the same, relatively small area. Several fishers talked of the problems caused by larger trawlers that cover a huge area and don't spend any money in the village or stay in the fishery long enough to care about the long-term viability of the stock. Creeling, on the other hand, tends to be done by skipper-owners with a small crew who all live and work in the community where the fishing ground is located. These kinds of embodied and spatial (and community) relations are important for the kinds of emotional attachments they form and their commitments to 'sustainable' fisheries.

11.3 Conclusion

What emerges from these issues is that the processes of boundary making are part of how emotional involvement and values towards particular aspects of fisheries management emerge. Such boundaries operate at several scales and levels of abstraction. For example, how people conceptualise the boundary between themselves and their environment is central to how they form notions of a realm 'outside' of themselves, which has implications for how they develop values towards and treat those environments. I see these boundaries as central to a sense of alienation from nature; an alienation that Milton (2002) has argued is at the heart of modern over-exploitation of resources. Or as the fisherman above put it, 'It's a business as opposed to a way of living.'

On a larger scale, how people conceptualise boundaries between themselves and other people is key to whether they will act for some notion of a collective good. In what contexts and in relation to which kinds of 'others' people will work with needs to be addressed in order to form effective groups to address environmental issues. Many people who work the land do have a sense of their resources and of belonging to a community, so why they choose to work together to protect their environment in some contexts but not in others is a question that needs to be addressed at the levels of subjectivity, community, politics and economics. The separation they see between themselves and 'environmentalists' can only be counter-productive to producing new practices that have ecological environments, communities and sustainability as key values. All of these factors are integral to the complex interplay between, and mutual constitution of, emotions, subjectivities and environments.

References

Allen, A. (2002). Power, subjectivity, and agency: Between Arendt and Foucault. *International Journal of Philosophical Studies, 10*(2), 131–149.
Bondi, L. (1993). Locating identity politics. In M. Keith & S. Pile (Eds.), *Place and the politics of identity* (pp. 84–101). New York: Routledge.
Bondi, L. (2005). Making connections and thinking through emotions: Between geography and psychotherapy. *Transactions of the Institute of British Geographers, 30*, 433–448.
Bondi, L., & Davidson, J. (2003). Troubling the place of gender. In K. Anderson, M. Domosh, S. Pile, & N. Thrift (Eds.), *Handbook of cultural geography* (pp. 325–344). London: Sage.
Brady, E. (2006). The aesthetics of agricultural landscapes and the relationship between humans and nature. *Ethics, Place and Environment: A Journal of Philosophy & Geography, 9*(1), 1–19.
Braun, B., & Castree, N. (1998). *Remaking reality: Nature at the millennium*. New York: Routledge.
Butler, J. (1997). *The psychic life of power*. Stanford: Stanford University Press.
Castree, N. (2005). *Nature*. London: Routledge.
Castree, N., & Braun, B. (Eds.). (2001). *Social nature: Theory, practice and politics*. Oxford: Blackwell.
Catchpole, T. L., Frid, C. L. J., & Gray, T. S. (2005). Discarding in the English north-east coast *Nephrops norvegicus* fishery: The role of social and environmental factors. *Fisheries Research, 72*, 45–54.
Clayton, S., Warden, L., & Opotow, S. (2003). *Identity and the natural environment: The psychological significance of nature*. Cambridge, MA: MIT Press.
Crean, K. (2000). Contrasting approaches to the management of common property resources: An institutional analysis of fisheries development strategies in Shetland and the Solomon islands. *Australian Geographer, 31*(3), 367–382.
Cronon, W. (1996a). The trouble with wilderness; or, getting back to the wrong nature. In W. Cronon (Ed.), *Uncommon ground: Rethinking the human place in nature* (pp. 69–90). New York: W.W. Norton and Company.
Cronon, W. (Ed.). (1996b). *Uncommon ground: Rethinking the human place in nature*. New York: W.W. Norton and Company.
Davidson, J. (2000). '…the world was getting smaller': Women, agoraphobia and bodily boundaries. *Area, 32*, 31–40.
Davidson, J. (2003). *Phobic geographies: The phenomenology and spatiality of identity*. Aldershot: Ashgate.
Foucault, M. (1991). Governmentality. In G. Burchell, C. Gordon, & P. Miller (Eds.), *The Foucault effect: Studies in governmentality* (pp. 87–104). Chicago: University of Chicago Press.
Gibson, K. (2001). Regional subjection and becoming. *Environment and Planning D: Society and Space, 19*, 639–667.
Gibson-Graham, J. (2002). Beyond global vs. local: Economic politics outside the binary frame. In M. Wright & A. Herod (Eds.), *Geographies of power: Placing scale* (pp. 328). Oxford: Blackwell.
Hanson, S., & Pratt, G. (1995). *Gender, work and space*. London: Routledge.
Haraway, D. J. (1997). *Modest_Witness@Second_Millennium. FemaleMan©_Meets_OncoMouse™*. New York: Routledge.
Henriques, J., Hollway, W., Urwin, C., Venn, C., & Walkerdine, V. (1984). *Changing the subject*. London: Methuen.
Hinchliffe, S. (2007). *Geographies of nature: Societies, environments, ecologies*. London: Sage Publications.
Little, P. D., & Watts, M. J. (1994). *Living under contract: Contract farming and Agrarian transformation in Sub-Saharan Africa*. Madison: The University of Wisconsin Press.
Longhurst, R. (2001). Geography and gender: Looking back, looking forward. *Progress in Human Geography, 25*(4), 641–648.

Longhurst, R. (2003). Introduction: Placing subjectivities, spaces and places. In K. Anderson, M. Domosh, S. Pile, & N. Thrift (Eds.), *Handbook of cultural geography* (pp. 282–289). London/New Delhi: Thousand Oaks/Sage Publications.

Mahoney, M. A., & Yngvesson, B. (1992). The construction of subjectivity and the paradox of resistance: Reintegrating feminist anthropology and psychology. *Signs, 18*(1), 44–73.

Malone, K. R. (2000). Subjectivity and the address to the other: A Lacanian view of some impasses in theory and psychology. *Theory and Psychology, 10*(1), 79–86.

Massey, D. (1994). *Space, place and gender.* Minneapolis: University of Minnesota Press.

McDowell, L. (1983). Towards an understanding of the gender division of urban space. *Environment and Planning D: Society and Space, 1*, 59–72.

Milton, K. (2002). *Loving nature: Towards an ecology of emotion.* London/New York: Routledge.

Nightingale, A. J. (2006). The nature of gender: Work, gender and environment. *Environment and Planning D: Society and Space, 24*(2), 165–185.

Nightingale, A. J. (2011a). Beyond design principles: Subjectivity, emotion and the (ir-)rational commons. *Society and Natural Resources, 24*(2), 119–132.

Nightingale, A. J. (2011b). Bounding difference: The embodied production of gender, caste and space. *Geoforum, 42*(2), 153–162.

Probyn, E. (2003). The spatial imperative of subjectivity. In K. Anderson, M. Domosh, S. Pile, & N. Thrift (Eds.), *Handbook of cultural geography*. London: Sage.

Ridgway, I. D., Taylor, A. C., Atkinson, R. J. A., Chang, E. S., & Neil, D. M. (2006). Impact of capture method and trawl duration on the health status of the Norway lobster, *Nephrops norvegicus*. *Journal of Experimental Marine Biology and Ecology, 339*, 135–147.

Rose, G. (1993). *Feminist and geography: The limits of geographical knowledge*. Oxford: Blackwell.

Scott, J. (1991). The evidence of experience. *Critical Inquiry, 17*(3 (Summer)), 773–797.

Scottish Executive (2005). *A strategic framework for inshore fisheries in Scotland.* Edinburgh: Scottish Executive.

Tuck, I. D., Chapman, C. J., & Atkinson, R. J. A. (2000). Population biology of the Norway lobster, *Nephrops norvegicus* (L.) in the Firth of Clyde, Scotland. II. Fecundity and size at onset of maturity. *ICES Journal of Marine Science, 57*, 1227–1239.

Whatmore, S. (2002). *Hybrid geographies*. London: Sage.

Wilson, N. (2001). Butler's corporeal politics: Matters of politicized abjection. *International Journal of Sexuality and Gender Studies, 6*(1/2), 109–119.

Chapter 12
Ethics and Aesthetics of Environmental Engagement

Nathalie Blanc

Abstract Current theories of justice cannot account for the motives of a sustainable environment, that is, the material perpetuation of a fair society (one that regards all its citizens as equals), because it ignores the extent to which ethics integrates fundamental care, and more particularly, moral virtues. Care ensures the perpetuation of an existing environment and the attention paid to its particularity, often at a local level (Paperman and Laugier 2005). Individuals and communities shape and nurture nature, while working with nature through gardening can contribute to wellbeing. This chapter explores such relationships through case studies of urban community gardens in Russia and the Netherlands, and considers how an ethics of care and aesthetic engagement can be integrated into human-environment relations. In Tver, a town between Moscow and St Petersburg, a group of scientists who wanted to organize an Agenda 21 were asked to assist some inhabitants of a block of flats adjacent to an industrial area. They formed an Ecological Club, and the garden they created played a key role in improving their quality of life. In Arnhem, a city in eastern Holland, close to the German border, the investment in urban public space by the inhabitants was the result of the withdrawal from the district by the authorities. During the 1970s, the inhabitants decided to fight against the degradation of their life-world and embarked on a project of gardens to achieve their aim.

Keywords Environmental justice • Environmental aesthetics • Local governance • Care

Current theories of justice cannot account for the motives of a sustainable environment, that is, the material perpetuation of a fair society (one that regards all its

N. Blanc (✉)
UMR LADYSS 7533, Université Paris 7-UFR GHSS,
Case 7001, 2 Place Jussieu, 75005 Paris, France
e-mail: nathali.blanc@wanadoo.fr

E. Brady and P. Phemister (eds.), *Human-Environment Relations: Transformative Values in Theory and Practice*, DOI 10.1007/978-94-007-2825-7_12,

citizens as equals) because it ignores the extent to which ethics integrates fundamental care, and more particularly moral virtues (for example, concern for automobile pollution). Care ensures the perpetuation of an existing environment and the attention paid to its particularity, often at a local level (Paperman and Laugier 2005). Furthermore, this concern for the 'local' is related to the idea of the global, which one's imagination can make fruitful.

Continuing this line of argument, how is it possible to characterize the relations between ethical and aesthetic engagement in public space? Approaching the question from the point of view of environmental care, which means taking care of individual beings (people and any living being and sometimes even non-living things), we can say that ethics and aesthetics are two sides of the same coin; the side that comes uppermost will be decided upon by the course of action according to the understanding of those involved.

From the start, we need to define more precisely the terms which are being used. The term 'environment' makes it possible to refer to the multiple links a person or any living being forges in its relations with the world. This singular organization adapts to its environment, making processes of creative adjustment; these processes can be described as an active environmentalization. They bring into play multiple sensory links, and this confers a certain aesthetic meaning to our living environments, making it significant and opening up the possibility of aesthetic judgments, which in turn may give rise to public debate. This is a pluralist version of environmental engagement. By 'aesthetic engagement', we do not refer to a specialized field which would be art, or to the philosophy of beauty, or even to a theory of taste. One must think of an active environmental mode of knowledge which is not reserved for art or culture.

This mode of active knowledge follows the reflections of John Dewey (1934) or, more recently, Arnold Berleant (1992) for whom aesthetic experience is a way of registering the environment inside oneself, instead of being the subject of a passive and disengaged contemplation. This is the reason why, moreover, Berleant doesn't agree with Kant's idea of disinterested aesthetic judgment. The aesthetic experience is a mode of training and a mode of knowledge which puts the body and the spirit to the test in single movement. Some forms of environmental mobilization, such as plural mobilizations within an urban framework aiming at the requalification of places of life, can be analyzed in terms of aesthetic engagement. Indeed, the categories mobilized (landscape, living space, ordinary patrimony, the beauty of gardens and floral spaces) to justify the movements, as well as the actors taking part in them (artists, landscape planners, etc.) are evidence of this.

In addition, these mobilizations testify to the plurality of significant and everyday links which give meaning to the environment. They give a role to images and imaginary compositions of places and an idea of what they should be, including directions on how to inhabit them, thus accounting for the environment in all its richness. The role of beliefs and emotions (including that of injustice), and that of the fictions that enable us to live together on a daily basis, should not be forgotten. Finally, by public space, we also designate the concrete formal space accessible to the public, as well as the virtual space of dialogue which indicates political debate. Public space is the political construction of a common world through debate and dialogue, but it is also a concrete reality, inasmuch as it is a living space. However, today, in the context of environmental complexity, public space is no longer simply the space of debate; it has become what's at stake in the debate. The environment as a social question, but also the natural and constructed environment, is turned into a tangible, shared and public space which brings forth a debate. Such a space is already contained in the law and contained in the public policies of landscape. Now that we have set the terms of the debate let us specify the places of ecological engagement.

We will proceed through three distinct steps: first, we will consider the extent to which the conceptual design of the environment is renewed by new media; a design which must include the ideas of both proximity and distance; secondly, we will discuss the question of environmental ethics in light of the concept of an ethics of care, as it has recently been developed in France by Patricia Paperman and Sandra Laugier (2005). A Russian example will show how the desire for a better life is guided by aesthetic judgment and experience.

Finally, we will discuss how such an approach upsets aesthetic prejudices. We will begin by pointing out the links between ethics and aesthetics, then how ethics based on self-knowledge goes hand in hand with an experience of the forms of life. A Dutch example will show how far this collective desire can accompany an urban transformation. In these cases, local and civil action is not politically instrumented: in Russia, the local state has given up management of the place; in the Netherlands, the inhabitants are allowed to invest themselves in local transformations.

12.1 An Environmental Conceptual Design Between Proximity and Distance

When Jacques Rancière (2000) speaks of 'sharing a sensitive world', he refers to the equal capacity of all members of a given society to decide what value to confer to things, things which are, in the environmental context, living and acting forces. Indeed, environmental problems endowed with an undeniable social effect are producing public policies and actions. One could even say that the emergence of environmental problems strongly contributes to legitimating civil society. Let us go further and ask: What did the environmental question produce in terms of new public objects?

Climate change, the development of all kinds of pollution, and its effects on health were addressed in autonomous fields of research independently of their social construction. The first stage of environmental issues in the 1970s, of sustainable development, focused on the ecological dimension of the relationship 'nature/ society'. This was mainly a question of ensuring that the spatial concentration and the activities of human populations do not further impact the environment, the latter being conceived on a strictly 'material' level (ecosystems, including landscapes, atmosphere, water, etc.) and analyzed in terms of the natural sciences which study the environment according to its various facets (ecology, physics, chemistry and their various combinations according to the studied objects). One can speak of an 'objective' approach to the environment. However, the environment cannot be reduced to its objective dimension.

What role does aesthetics play in the definition and co-construction of a new public space which incorporates new objects, those speaking to the senses and affect? What role can aesthetics play to overcome the neglect of significant ordinary links to the environment? Precisely, its role is to initiate and inform public discussion concerning a new 'shared sensibility', implying both sense perception and sensible 'behaviour'. What creates such 'sensibility' over and above what is collectively defined as 'sensible' are difficult neighbourhoods, fragile or irremediably corrupted environments, endangered species, and so on.

However, we should specify why we think that local space, which integrates references to more remote spaces, is to us a particular space of investment. Public discussion is necessary, inasmuch as it relates to its objects as objects endowed with a material and social nature, that is to say, as objects that are the products of interactions between nature and society. This public space of proximity is one thread of the knot which constitutes the construction of an environmental conscience. In our virtual world, it remains necessary to construct and enrich the discussion by making precise reference to specific places of dwelling, of life. These living environments give to each of us the possibility of forging means of acting, of building a representation of the world – which pushes us to work out a point of view on the environment, its necessary safeguarding as well as the means of adapting it to ecological upheavals in a creative way.

Thus if we want aesthetic debate and the question of creativity to reach beyond the field of art, we are compelled to produce a new horizon of action: environmental aesthetics is a field where one can sketch a role for aesthetics in policy-making.

Why are we concerned about our local environments? Aesthetic discussion opens up a space in which to construct an imaginative framework of reference, a representation of the significant space in which we live and also to which we aspire. It is a question of being equipped with tools which do not reduce human societies to a mass subjected to the predicted catastrophe of climate change. It is a question of inventing the means of working out answers which do not consist only of forced adaptation to an endangered ecology, but which also represent nature in new ways and thereby suggest new forms of collective action.

Such a concept of local action is necessarily linked to a concept of global action: the local is not simply a question of geography, of a more or less measurable space associated with a given group or society, it is also a question of the multiplicity of emotions and relational networks that the local environment engenders. Indeed, it is necessary to revise the question of the local: we should make a place for new means of communication, but also for the imagination. They both transform distant relationships into close ones; they permit us to include remote places in our hearts and to care for them. They are made of all these picturesque accounts that one gives oneself, and contribute to the texture of the place we live in. Seen in this way, fiction helps to reconsider the how and the where of our experience of life and is thus special in terms of activity.

If we want to try to account for the various types of relationship between proximity and distance, the local and the global, we must take aesthetics into account. The local is what we think of as a relation of proximity and what we hold to be ours.

12.2 A Different Model of Ethics

We will conceive of ethics in terms that differ from those conventionally used. A strict definition of ethics would consider it as only a principle of self-regulation of self-knowledge. As Agamben explains, ethics is the sphere which knows neither fault nor responsibility, it is the doctrine of the happy life, 'To recognize a fault and a responsibility amounts to leaving the sphere of ethics in order to penetrate into that of the law' (Agamben 2003: 25). How could this lived ethics, conceived as the self-realization of the individual, touch on concern for the environment? The Kantian universalist concept of morals or the utilitarianism of Mill are among the two principal traditions in moral philosophy. Kantian moral theory argues that moral commands must take the form of universal and applicable laws (i.e., laws that apply to any rational being placed in a similar situation) independently of the individual characteristics (such as desires or interests) that motivate a given action. Classical utilitarian ethics postulates that the only common point among people is their aspiration to happiness: the hedonist principle. Both of these moral theories begin with the fact that human beings share the same principle of reason or the same capacities and desires for pleasure. The former recommends considering ourselves and our particularities in the abstract; the latter founds its approach on the relation of self-interest which controls individuals. Recent work in ethics renews the debate in an interesting way with an 'ethics of care', and thus an ethics based on feelings of 'care', and more particularly on the concerns of close relationships and proximity, an attention focused on the singular. For example, according to Ambroise,

> Carol Gilligan identifies three fundamental characteristics differentiating ethics of care from an ethics of justice. First, the ethics of care is articulated around moral concepts different from the ethics of justice, namely, human responsibility and its connections rather than rules and the law. Second, this form of morals is related to concrete circumstances and is not formal and abstract. Third, this form of morals is better expressed not as a set of principles and rules but as an activity and a practice, 'the activity of care'. Thus, according to Gilligan's' 'different voice', morals are not founded on abstract and universal principles, but in the experiences and moral problems that ordinary people meet in daily life. In short, the central moral principle consists of 'care', in not harming others and avoiding selfishness' (Ambroise 2005: 264).

Our recent research studying the individual's investment in their lifeworlds shows the extent to which such a concept of ethics may resonate with environmental concerns conceived in particular in terms of relations of proximity. Can the investment of people in their lifeworlds go hand in hand with an ethics of the care?

The ethics of care does not distinguish happiness and moral virtue as strongly as Kant does. For Kant, morality and happiness are distinct things: the former is controlled by the sense of duty, while the latter is driven by self-interest or by particular inclinations. Seen through the eyes of the philosophers of care, the purpose of ethics is not to determine which moral principles form rules for action. Rather, ethics enables us to wonder about the life we want to lead and the person that we want to be or become. This is why care ethics is understood on the level of the actor, and in particular his or her emotions, his or her happiness and values, and the web of relationships within which they are situated (rather than at the level of action). In this way, happiness is not only pleasure, but corresponds to the adequacy of individual values and their recognition in the full sense of the term, i.e., the place they have taken in the world.

What does such an approach contribute to an ethics of environment? Let us say that it is not just a question of integrating into the rules the individual sets for herself a recognition of her environmental impact (and the need to address it). The environment as a social question modifies the question of ethics more substantially. It invites each one of us to become aware of the danger to our bodies and, more generally, to human lives due to the impact of human activities on the environment. It invites each one of us to reflect on the conditions of our adaptation to the economic and social context we move in. To adapt is also to maintain the conditions of one's individual happiness, which presupposes being open to environmental opportunities. Furthermore, adapting means deriving from environmental knowledge the terms according to which it will be possible for us to create a happy life, i.e., the various factors must enter into agreement in such a way as to produce the internally desired effects. It is a question of anticipating, of imagining the impacts of one's actions on concrete reality (and vice versa). Aesthetic readings of what surrounds us, and the perceptual challenges which affect the subject facing change and weaken her will to change her living conditions, build a concern around the preliminary and corollary conditions of action. They constitute affection or, at least, an attentive point of view which is the prerequisite of care, and which dramatically changes our sense of responsibility and our agency. This concern is a mobile activity, a mobility fuelled by perception as much as by the intellect. Therefore, there cannot be any discussion of ethics which ignores the environment, taken in the general sense of what surrounds us or even in the more specific sense of what ecology defines as the environment. My wellbeing is identified in particular with the protection of this irreplaceable singularity, a relation that is symbolised by the beautiful object (or beautiful environment) (Breviglieri and Trom 2003).

The value of my wellbeing is also identified with the idea of participating in the construction of a common project of resistance to public policies or human behaviours which are destructive. Aesthetics appears in this case like an indicator of wellbeing. Indeed, aesthetics corresponds to the construction of an ethics through shaping facts successfully (a co-evolution). Environmental aesthetics is part of happy ethics. In this way, the key action is the one that exemplifies the good of my wellbeing in relation to good living conditions.

12.3 Environmental Action in Tver

This section discusses the opinions of the women of Tver,[1] a town between Moscow and St Petersburg, who were interviewed for this research. A group of scientists who wanted to organize an Agenda 21[2] were asked to assist some inhabitants of a block of flats adjacent to an industrial area. The inhabitants formed an Ecological Club, and the garden which they created played a key role in improving their quality of life. More specifically, as the women who started this environmental action said themselves, it enhanced their everyday life: *'They wanted to make it beautiful, as beautiful as elsewhere, as in the botanical garden'* (Pushai, May 24, 2004). The goal of the Ecological Club was to create a corner of nature and a place of recreation which offered comfort comparable with the botanical garden of the old city: *'It is possible to have an environment as beautiful everywhere, not only here'* (Pushai, 24/5/04). It is necessary to emphasize the importance of parks in the life of Russian city dwellers. Indeed, the revaluing of the lived environment is undertaken in many national contexts, but it is even more striking in Russia, due to the way parks are valued and the existence of gardens with flowers in particular. The garden (which can be limited to a few flowers), is a space of personal and collective realization, demonstrating know-how, a place where relations with nature are paramount.[3] The scientists and the association of the botanical garden provide advice and thus help to plant the seeds of a progressive regeneration of this district. They make this project a major element of a green frame woven out of elements of local biodiversity. This fits into the broader context of a civic environmentalism that has begun to spread and that is supported by inhabitants forming networks of local associative action (Blanc and Emelianoff 2007). This experiment in the course of which plants

---

[1] Interdisciplinary program of research on urban sustainable development PIDUD-CNRS/SHS, Blanc, N., Emelianoff, C. (2005). Carried out in partnership with the Institute of Sociology of the Academy of Science of Moscow, the study related to new forms of environmental mobilization in four large cities, including Bryansk and Perm.

[2] Agenda 21 is a plan of action for the twenty-first century adopted during the Earth Summit in Rio held in 1992. The territorial collectivities are called on in the framework of chapter 28 of Agenda 21, to put into place a program of Agenda 21 at their scale, integrating principles of sustainable development through a population consultation mechanism: thus was born the local Agenda 21 'they play, on an administrative level closer to the population, an essential role in education, mobilisation and taking into account the views of the public in favour of sustainable development' (Chapter 28).

[3] The relationship with nature is inherited from the pagan pre-orthodox culture: a great nature, a fusion nature with limits one does not perceive. This is cultural, making Russian earth and nature a 'matrix', and it is found in both ordinary practices and the sensitivity of the ecologist. It impregnates Russian ecology considerably and plays a part in the scientific qualification of the earth as biosphere; animating the close connections to nature highlighted within the field of the protection of humans and especially of territory. Russian nature, with all its symbolism, plays a role in territorial unification: it is identified as a common substrate. Nature, immense, merges with the earth and the fatherland. The frequent use of the term 'Russian nature' testifies to the indistinguishability of the concept of nature and territorial identity, which is not conceivable in France, for example. Which meaning indeed would one have of 'French nature'?

are used to remake the bases of buildings has led to a progressive assumption of responsibility by the inhabitants, of care for the communal parts and the external spaces, a responsibility that had been abandoned by the housing office. The initial plan consisted of the creation of a garden in 2001; a playground and other games for children were realized and repair of the staircases carried out. This project concerns an extremely degraded environment whose abandonment by the authorities obliged the inhabitants to take over. The project comprises mostly women, although some children and teenagers are also involved. It employs the vice-director of the school, as well as a few of the owner-occupiers of the residences, although the wages offered are relatively small. It is also necessary to note the important role given to such aspects as garden internal and external decoration. Certain inhabitants carve wooden statues which come to ornament the gardens, as in the parks of large cities. This contributes to the construction of a collective heritage, which may have a significant impact on visitors (among whom we count the researchers) (Stavo-Debauge and Trom 2004). This case study shows the importance of the argument of beauty in the improvement of one's local habitat. It is a space designed like a natural environment. The scientists and the local ecologists present on the spot understood that their interests coincided with those of the inhabitants. For all, the stake was to plant local species in order to improve urban public space. For the scientists and the Ecological Club, it was a question of contributing to local biodiversity; for the city dwellers, to find nature in their daily living space. What is demonstrated here is the problem of facing up to environmental degradation as it is represented and objectified on a scientific level in the name of a better life, which involves the enhancement of an area and, more generally, an appreciation of the beauty of nature. Taking into account the beauty of nature as an important part of concern for environmental preservation in our period of global networks may be a way of developing the concept of care and attention to what surrounds us. This is a principle of justice which must be extended and associated with the responsibility that it creates. It is important to show that non-assistance for an endangered environment may influence individual behaviour. Indeed, the impact of human activity on the environment narrows the possible field of individual and collective adaptation. Creating awareness of this relation amounts to emphasizing the feeling of responsibility towards the environment as a form of responsibility for the future.

12.4 Ethics and Aesthetics with Respect to the Environment

Thus the aesthetic may be seen as a mode of knowledge and transmission of what is right in the environment, allowing the best kind of adaptation or recognition of the values shared by individuals and the collective. It can define the way in which we justify the irreplaceable quality of our environments: the idea of a heritage associated with a landscape, the modes of asserting the identity of a place – these are ways of accounting for a natural and singular irreducibility. Disorder and damage in the environment provides the occasion for these values to appear. Consequently, the aesthetics of the environment is not just the recognition of an aesthetized appearance, that is to say, qualities which make it possible to describe the beauty or the ugliness of a place. Nor is it simply a method of knowing a place. It is, more precisely, the domain where taste is formed, where taste can then be shared and communicated. It could be described as the process of learning about a shared environment, based on habitability in constant negotiation. Defined thus, aesthetic judgment is taken into a course of action, in contradistinction to other types of judgment. It is knowledge acquired through action and not derived from an overarching point of view. As Kant (2000) wrote in 1790, the experience of the fine arts, by which humanity acquires some capacity to communicate on a very intimate but also on a universal level, can teach people to live together. Further on, after having indicated that people governed by the law develop an aesthetic sense from the reciprocal exchange of ideas about the judgment of taste, Kant insists on the fact that the future, which will move away from nature, obliges us to go in this direction, since it will be necessary for people to learn how to evaluate their action by joining together 'legal constraints required by the highest culture with the force and the accuracy of free nature feeling its own value' (1995: 344). Obviously such a development implies, as Kant adds, the development of moral judgment, since taste is strongly dependent on it. In what way? Refining our taste is also an opening in terms of spiritual and moral elevation. Ultimately, it is not simply a question of moral duty, which may be contradictory to the spirit of happiness, but of self-knowledge, and of one's inclinations, which lead to happiness. Indeed, the exact idea of such happiness corresponds to that of the moral sense, which is the horizon of happiness (of any happiness) when one thinks of the relations between causes and effects in a given society.

What we will show now is that giving shape to a district via gardens and the activity of gardening and, overall, placing it in relation to a certain urban landscape, is accompanied by an ethical will which corresponds to the specific idea of a happy life in a given place and urban context. Such a space comes into being gradually: the possibility of action engendered in the course of and due to environmental action creates new horizons incrementally. In Arnhem, a city in eastern Holland, close to the German border, the investment in urban public space by the inhabitants was the result of the withdrawal from the district by the authorities. During the 1970s, the inhabitants decided to fight against the degradation of their lifeworld and decided on a project of gardens in order to achieve their aim. During this period, one of the inhabitants (Gerda Van Steijn, who looks after one of the gardens today) decided to make use of large flower pots on balconies and to invite other inhabitants of this district to plant flowers in them: 'they do not want bad people to invade their district' (Discussion with Loet Van Moll, 17/06/2007). According to the story now told by one of the principal actors in the history of these gardens, the epic of this laundress gave the impetus to what constitutes, in a structured and visible way in terms of urban space, the organization by the 'Groengroep' of eleven gardens[4] in a district which mixes

[4] These gardens are called: Schooltuin, Kruidentuin, Ecologische tuin, Watertuin, Emmeplantsoen, Psttuin, Bestedelingentuin, Achtertuin, Spoortuin, Boembakkenactie et Straatgroen. The internet site of the 'Groengroep' is: www.spijkerkwartier.net/wegwijs/groen. Accessed 1/6/11.

social habitat, middle class people and populations with high incomes. Indeed, it is important to note that gentrification accompanied the regeneration of this district.

This first action is still remembered as crucial today; the women moved from flower pots on balconies to planting trees, and the majority of the trees of the district go back to this time. Then, in 1983, the 'Groengroep' decided , upon the initiative of a professor of biology, to create an ecological garden, and they needed the help of the municipality to do it. The conflicts were numerous and each garden tends to reflect the aesthetics and organizing principles of its creators. These differences do not seem to be a problem, though: the different gardens and the process leading to their creation can be seen as a whole. The ecological garden was completed in 1986 and includes native plants of the area, fountains and small bowers resembling the 'follies' of the nineteenth century. Artists took part in the construction of this 'urban jewel' surrounded by a group of buildings. Such a socio-spacial design, assuring total visibility and thereby allowing social control, makes it possible for children to play there alone. There are 15 people today who are in charge of taking care of the place. Gradually other inhabitants became involved in the initiative, partly determined to fight against municipal attempts to replace the last vacant lots and parks. We can therefore note that such an initiative is also a political action; greenness is synonymous with resistance in many ways. Today, more than 50 people in the district are involved in the gardens and a hundred people garden for a few hours per week. The 'Groengroep' holds excursions, guided tours, offers natural spaces for schools, and there are more than one hundred visitors per year. In addition, looking after these gardens has become a local tradition with meetings four times per year because the inhabitants wish to preserve a strong link among themselves. Once a year, thanks to the money distributed by the municipality, the 'Groengroep' orders plants.

The relations between the local authorities and the inhabitants are now based on confidence, the reason for this being the municipality's having realized that 'it works to work with those who go forward…'. The gardens are varied: Watertuin, the watergarden opened in 1993, is established in the place of the former Gelders Spijker ponds. This inspired the inhabitants to build a garden with large ponds which shelter frogs, salamanders and where herons even some to fish. Achtertuin was developed on the site of an old barn which was used to store grain (hence the name Spijker), and was replaced by the workshops of craftsmen. In the 1900s there remained nothing of these workshops except one structure used to store coal. After its demolition in 1985 a garden which was located between all these buildings developed according to the desires of the inhabitants.

This living initiative is based on a very effective urban policy regarding public space and parks. Indeed, the city planning for the period 2003–2015 is resolutely green; it is founded on the regeneration of public space and based on the relations between plants and water. The projects work on both a horizontal and a vertical level; the idea to plant the fronts of buildings is done on the initiative of the inhabitants. This regeneration of public space by means of plants and the revaluing of open spaces sometimes comes into conflict with the economic interests of the city as well as with the need to build denser neighbourhoods to address the requirements of sustainable cities. This is all the more true as large companies, such as the insurance company

Axa, have chosen to settle in the city. Moreover, this urban project takes into account the need for a restoration of the sewers, in particular by the separation of rainwater and waste water. The gardens and porous grounds will assume the role of tanks for rain water which will percolate down into the ground. This project was the subject of a consultation with the inhabitants and the 'green plans' were discussed district by district. The city of Arnhem counts 25 districts and 12 of them have the equivalent of local consultancy boards. This discussion is part of a long Dutch tradition of participation which consists in thinking 'that it is necessary to speak until finding a solution'. Contracts are signed with the inhabitants in order to insure themselves against bad management or to protect them from the risks of lawsuit. Spaces of free gardens are provided by the municipality as well as the budget for managing the gardens. For example, the inhabitants of the district of 'Spijkerkwartier/Spoorhoek' receive 5,000 euros per year via the 'Groengroep', which enables them to assume the management of these gardens (seeds, tools, etc.) These groups or associations, although their legal status is not always formal, are clearly informed of their rights and duties with regard to the municipality and fall under the green general policy of Arnhem.

Today there are six groups equivalent to that of 'Spijkerkwartier/Spoorhoek' which are entirely based on the participation of volunteers. The idea which dominates the policy of dialogue is to keep contact (good contacts) with the population and in this way to preserve the future, i.e., to make possible great transformations when that proves necessary. Such a state of mind helps to build a profitable capital of 'mutual confidence'. There are no measurements of this capital primarily deriving from communication and human relations. Basically, the policy of Arnhem is based on the idea of ensuring the best service for citizens. Two points are important here. First, the local actors were unaware of the terms of 'biodiversity', or of the question of sustainable development such as it was stated at the Summit of Rio in 1992, but that did not prevent strong ecological investment. Second, from the point of view of aesthetics, such an investment is paramount. These communal actions add up to a considerable result; these gardens, which we visited one after the other, are small marvels of variety in their aesthetic and floral aspects. When one thinks of the effort made to achieve such a network of gardens over more than 20 years (which, in addition, seems to have considerably increased local biodiversity, if one believes the observations of the gardeners and inhabitants), the result is certainly remarkable. Why such a powerful representation of green places and gardens in the regeneration of space? The answer is of course complex, fusing various aspects.

The garden represents a haven of peace in urban life which one closely associates with Paradise; it is a place of reading, relaxation, possibly prayer, and of play too. It is a place where one finds oneself assured of peaceful social contact or, on the contrary, a place of refuge for marginal people and social outsiders. It is also a symbolic space which encourages the alliance of the inhabitants and at least partly ensures social links. These gardens go against a theoretical and abstract environment, since they confront us with living beings, growing, proliferating, multiplying.[5]

[5] In 'Elegy for a Garden', Andrew Light (2004) agrees with the idea that a complete environmental ethic ought to include all environments, including the more cultivated green spaces of urban environments.

The investment of these spaces is incontestably aesthetic where the initial motivation for building them is concerned, but also in terms of the spectacle presented to city dwellers. The gardens are discrete decorations, built into the urban fabric, of rich scenes and situations, but this is also an ethical investment which corresponds to the hope of a better life in the various meanings of the term: collective and individual. Form guides the action in this project designed to lead to a better life.

It is important to encourage the aesthetic participation of the inhabitants, but it is just as important to enrich the social and environmental analysis of a different aspect, that is, sensitivity to the environment. It is indeed urgent to develop a true art of the environment which engages aesthetics and ethics. If it is true that the sense of the word environment is more complete when its aesthetic dimensions are taken into account, then art, in so far as it may be said to constitute the avant-garde of future perception and representation, can contribute to that reflection: the environment is an art, an eminently social art, collective art, the art of the fabric of life. It is the responsibility of all, and not only of women, who often manage local environments,[6] to develop a constructive relation to the environment. And that requires a quite particular kind of teaching, because the judgment of taste is the art of learning how to judge by oneself. We live in a period where we can observe the 'failure' of political dogmas, and thus we are required to develop new vectors of autonomy. Seen in this way, urban art is autonomy; being detached from nature implies having to learn about oneself.

References

Agamben, G. (2003). *Ce qui reste d'Auschwitz*. Paris: Payot Rivages.

Ambroise, B. (2005). Réalisme moral et éthique du care. In P. Paperman & S. Laugier (Eds.), *Le souci des autres, Éthique et politique du Care* (pp. 263–278). Paris: Éditions de l'École des Hautes Études en Sciences Sociales (EHESS).

Berleant, A. (1992). *The aesthetics of environment*. Philadelphia: Temple University Press.

Blanc, N., & Émelianoff, C. (2005). *La ville durable face à l'écologie des pratiques: Politiques et cultures de l'environnement urbain dans les villes russes*. Paris: Programme interdisciplinaire développement urbain durable, CNRS/SHS.

Blanc, N., & Émelianoff, C. (2007). Réappropriation du sol et société civile dans deux villes russes. *Environnement urbain/Urban Environment, 1*, 31–41. www.vrm.ca/cyber-revues.asp Accessed 1 June 2011.

Breviglieri, M., & Trom, D. (2003). Troubles et tensions en milieu urbain. Les épreuves citadines et habitants de la ville. In D. Cefai & D. Pasquier (Eds.), *Le sens du public: Publics politiques et médiatiques* (pp. 399–416). Paris: PUF.

Dewey, J. (1980) (1934). *Art as experience*. New York: Perigee Books.

[6] It should be noted that feminist theories within the ethics of care can be partly addressed by drawing on French political ecology. This refusal to be reinterrogated relates in particular to researchers in the social sciences concerned with empirical work, where the great presence of women is found in the management of proximity of the environment.

Kant, I. (2000). *Critique de la faculté de juger* (1e éd. 1787), Traduction, présentation, bibliographie et chronologie par Alain Renaut. Paris: Garnier-Flammarion.
Light, A. 2004. 'Elegy for a Garden'. http://www.terrain.org/essays/13/light.htm. Accessed 1 June 2011.
Paperman, P., & Laugier, S. (Eds.). (2005). *Le souci des autres: Éthique et politique du Care*. Paris: Éditions de l'École des Hautes Études en Sciences Sociales (EHESS).
Rancière, J. (2000). *Le partage du sensible: Esthétique et politique*. Paris: La Fabrique.
Stavo-Debauge, J., & Trom, D. (2004). Le pragmatisme et son public à l'épreuve du terrain: Penser avec Dewey contre Dewey. In B. Karsenti & L. Queré (Eds.), *La croyance et l'enquête, aux sources du pragmatisme* (Raisons Pratiques, Vol. 15). Paris: Éditions de l'École des Hautes Études en Sciences Sociales (EHESS).

Index

E. Brady and P. Phemister (eds.), *Human-Environment Relations: Transformative Values in Theory and Practice*, DOI 10.1007/978-94-007-2825-7,

MIX
Papier aus verantwortungsvollen Quellen
Paper from responsible sources
FSC® C105338

If you have any concerns about our products, you can contact us on
ProductSafety@springernature.com

In case Publisher is established outside the EU, the EU authorized representative is:
Springer Nature Customer Service Center GmbH
Europaplatz 3, 69115 Heidelberg, Germany

Printed by Libri Plureos GmbH
in Hamburg, Germany